Threatened Birds of Jammu & Kashmir

Threatened Birds of Jammu & Kashmir

Asad R. Rahmani, Intesar Suhail, Pankaj Chandan,
Khursheed Ahmad, and Ashfaq Ahmed Zarri

Maps prepared by

Mohit Kalra and Noor I. Khan

Layout and design by

V. Gopi Naidu

Sponsored by

Pavilion Foundation, Singapore

Oxford University Press, Walton Street, Oxford OX2 6DP
Oxford, New York,
Athens, Auckland, Bangkok,
Cape Town, Chennai, Dar-es-Salaam,
Delhi, Florence, Hong Kong, Istanbul, Karachi,
Kolkata, Kuala Lumpur, Madrid, Melbourne,
Mexico City, Mumbai, Nairobi, Paris,
Singapore, Taipei, Tokyo, Toronto,
and associated companies in
Berlin, Ibadan

Recommended citation: Rahmani, A.R., Intesar Suhail, Pankaj Chandan, Khursheed Ahmad, and Ashfaq Ahmed Zarri (2013) *Threatened Birds of Jammu & Kashmir.* Indian Bird Conservation Network, Bombay Natural History Society, Royal Society for the Protection of Birds, and BirdLife International. Oxford University Press. Pp. xiv + 150.

Consultant Editor: Gayatri W. Ugra

Layout and design: V. Gopi Naidu

Maps: Mohit Kalra and Noor I. Khan

IBCN, c/o BNHS, Hornbill House, Shaheed Bhagat Singh Road, Mumbai 400 001, India
Telephone: 0091-22-22821811, Fax: 0091-22-22837615
Email: ibabnhs@gmail.com and info@bnhs.org
Websites: <www.ibcn.in> <www.bnhs.org>

Bombay Natural History Society is registered in India under the Bombay Public Trust Act 1950: F244 (Bom) dated July 6, 1953

ISBN :   9780199452699

Proceeds from the sale of this book will go to the Indian Bird Conservation Network

Front Cover : Black-necked Crane by Dhritiman Mukherjee
Back Cover : Kashmir Flycatcher by Clement Francis

Available from
IBCN, c/o BNHS, Hombill House, Shaheed Bhagat Singh Road, Mumbai 400 001, India
Telephone: 0091-22-22821811, Fax: 0091-22-22837615
Email: ibabnhs@gmail.com and info@bnhs.org
Websites: <www.ibcn.in> <www.bnhs.org>

Processed by Trendz Phototypesetters. Email: gotrendz@gmail.com
Printed by Specific Assignments India Pvt. Ltd. Email: info@specificassignments.com

CONTENTS

CRITICALLY ENDANGERED

ENDANGERED

VULNERABLE

NEAR THREATENED

CONTENTS (*contd.*)

MARGINAL

REFERENCES

■ CRITICALLY ENDANGERED   ■ VULNERABLE

■ ENDANGERED   ■ NEAR THREATENED

PREFACE

Although wildlife protection is on the concurrent list of the Government of India, most of the conservation action takes place at the state level. The role of state forest and wildlife departments, general public, local communities, national and local NGOs, and lay citizens is immense.

The first author of this book brought out a major tome in 2012 titled *Threatened Birds of India* which was a supplement to the work of BirdLife International at the global level. Soon, a need was felt for state-wise smaller books which would be more useful, as most of the conservation action is decided and implemented at state level. The first book in this series was *Threatened Birds of Assam* (2012), which was soon sold out. In 2013, the second book in this series *Threatened Birds of Uttarakhand* came out, soon to be followed by *Threatened Birds of Maharashtra*. This is the fourth book in this series.

The present book follows the list of globally threatened and near threatened bird species published by BirdLife International in 2013. BirdLife regularly updates the list for International Union for Conservation of Nature (IUCN). Thus, this book provides comprehensive information about two Critically Endangered species, two Endangered species, 11 Vulnerable species, and nine Near Threatened species which have been reported from Jammu & Kashmir. Some marginal species of various categories have also been described in brief.

Besides the specific recommendations for the conservation of each threatened species found in Jammu & Kashmir, we have also given general recommendations for protection of all bird species and their research needs. The book provides distribution maps of the important threatened species. The book will be useful to government agencies like the Forest Department, Wildlife Department, conservationists, researchers, and birdwatchers alike.

By protecting threatened birds, we will be able to save their habitats as well. The habitats of birds are home to innumerable other species of flora and fauna, which will also derive the benefit of our conservation initiatives.

Authors

A.K. Singh, IFS
Principal Chief Conservator of Forests (WL)/
Chief Wildlife Warden
Jammu & Kashmir Government

FOREWORD

I am delighted to know that the Bombay Natural History Society (BNHS) is bringing out a book *Threatened Birds of Jammu & Kashmir*. The book is published in collaboration with Bird Life International and the Royal Society for the Protection of Birds, both based in United Kingdom. It will be distributed by Oxford University Press.

Jammu & Kashmir has rich biodiversity spread over all the state, but particularly in the Kashmir Valley and Ladakh.

It is important to know that Jammu & Kashmir has three out of 15 Critically Endangered bird species of India. Apart from this, there are 11 Vulnerable and 20 Near Threatened species. Specific recommendations given in the book will help in protecting these birds. We need to take effective measures so that all these species and even our common species are protected.

Jammu & Kashmir has 25 Important Bird Areas (IBAs) and seven more have been recognised recently. These IBAs are extremely important for the protection of threatened birds.

The Bombay Natural History Society and especially Dr. Asad R. Rahmani, the Director, has been very considerate with the state of Jammu & Kashmir, helping the state with regard to research in birds and helping us in various aspects of their conservation. The recent efforts by him in collaboration with WWF-India and WII in ringing, collaring, and fixing of PTT on Bar-headed Goose, Black-necked Crane in Ladakh during autumn of 2013 and such an exercise in Gharana Wetland during February 2014 in Jammu, are examples of his personal efforts and involvement. It will enrich our understanding of bird migration in the state and the use of habitats by birds to a large extent.

I congratulate the authors of the book, Dr. Rahmani and others, for bringing out such a valuable resource on the birds of the State of Jammu & Kashmir. With support from organizations like WWF, BNHS, WII, and WTI, I am sure that we will be able to protect various species listed in this book.

I hope the book will be used by decision makers, researchers, NGOs, conservationists, and members of civil society. We need more such books on different taxa to appreciate and protect the rich biodiversity that our state possesses.

A.K. Singh, IFS

Asif M Sagar, IFS
Chief Conservator of Forests (Wildlife)
Jammu Region
Jammu

MESSAGE

The State of Jammu & Kashmir is bestowed with a wide range of habitats and ecosystem types ranging from low lands of Jammu to Trans-Himalayan region of Ladakh. Jammu & Kashmir is well known for its unique assemblage of avifauna including various migratory as well as resident bird species. Out of the 1,300 species of birds found in India about 500 are found in J&K. There is an urgent need to protect this unique natural heritage.

I am delighted to note that a very important book 'Threatened Birds of Jammu and Kashmir' is being brought out by Bombay Natural History Society. I congratulate Dr. Asad R. Rahmani and other authors for their dedicated efforts to bring out this really commendable book.

Despite the best and sincere efforts from wildlife managers, researchers, policy makers, NGOs and others concerned with this subject, threats to these bird species have increased over the years. Moreover, the information related to birds and research in this field continues to be inadequate and this poses difficulties in taking effective decisions for their conservation. In this context I find this book to be of great help to decision makers and policy makers in making assessment of the actual ground-level situation and taking appropriate measures accordingly. I am sure this book will be a handy tool for the researchers and conservationists working on the birds of Jammu & Kashmir.

It gives me great pleasure to be part of this useful publication for our state. I hope this book will bring the readers closer to nature and lend them inspiration to help conserve various threatened birds of the state.

Asif M Sagar, IFS

ACKNOWLEDGEMENTS

The authors wish to thank the Pavilion Foundation, Singapore for providing financial support for the publication of this book. We also thank Ms Meileen Choo, Executive Director, Cathay Organisation Holding Ltd., Singapore.

In BNHS, we want to thank Dr. Gayatri Ugra for editing the text, Mr. Gopi Naidu for layout and design, and Mr. Mohit Kalra and Mr. Noor Khan for preparing the maps.

Our gratitude to Mr. Homi Khusrokhan, President, BNHS; Mrs. Sumaira Abdulali, Hon. Secretary; Mr. E.A. Kshirsagar, Hon. Treasurer; and all members of the Governing Council for their support.

Among our colleagues at BNHS, we would like to thank Mr. Abhijit Malekar, Mr. Atul Sathe, Dr. Deepak Apte, Ms Divya Varier, Mr. Divyesh Parikh, Mr. Isaac Kehimkar, Mr. J.P.K. Menon, Mr. M.G. Mathews, Mr. Mrugank Prabhu, Ms Neha Sinha, Ms Nirmala Barure, Ms Nisha Shilaj, Ms Parveen Shaikh, Mr. Rahul Khot, Mr. Sachin Kulkarni, Mr. Sameer Bajaru, Mr. Santosh Mhapsekar, Mr. Siddhesh Surve, Ms Sonali P. Vadhavkar, Dr. Swapna Prabhu, Mr. Tarendra Singh, Ms Tejashree Nakashe, Dr. V. Shubhalaxmi, Ms Varsha Chalke, and Ms Vibhuti Dedhia.

We would like to thank the following persons from the State of Jammu & Kashmir: Mr. A.K Singh, IFS, Principal Chief Conservator of Forests (Wildlife), Jammu & Kashmir; Mr. A.R. Wani, former Chief Wildlife Warden, Jammu & Kashmir; Mr. A.R. Wadoo, former Chief Wildlife Warden, Jammu & Kashmir; Mr. O.P Sharma, IFS, Chief Conservator of Forests (Ecotourism); Mr. Jigmet Takpa, IFS, Conservator of Forests (Wildlife), Ladakh Region; Mr. Asif M. Sagar, IFS, Conservator of Forests (Wildlife), Jammu Region; Mr. M.S. Bacha, former Regional Wildlife Warden, Dept. of Wildlife Protection; Mr. Naseer Kichloo, former Regional Wildlife Warden, Dept. of Wildlife Protection; Mr. Mushtaq Ahmad Parsa, former Wildlife Warden, Dept. of Wildlife Protection; Mr. Rashid Yahya Naqash, Wildlife Warden, Dept. of Wildlife Protection; Mr. A.B. Rouf Zargar, Wildlife Warden, Dept. of Wildlife Protection; Mr. Imtiyaz Ahmad Lone, Wildlife Warden, Dept. of Wildlife Protection; Mr. Tahir Shawl, Wildlife Warden, Dept. of Wildlife Protection; Ms Ifshan Dewan, Wildlife Warden, Dept. of Wildlife Protection; Mr. Mohammad Maqbool Baba, Wildlife Warden, Dept. of Wildlife Protection; Mr. Mohammad Sadiq, Wildlife Warden, Dept. of Wildlife Protection; Ms Sameena Ameen, Research Officer, Dept. of Wildlife Protection; Dr. Junaid Nazir Shah, Mr. Shahid Bashir, Dr. Rahul Kaul, Wildlife Trust of India; Mr. Riyaz Ahmad, Project Lead J&K, Wildlife Trust of India; Mr. Pushpinder Singh Jamwal, WWF, Leh; Mr. Tassaduq Mueen, Birdwatcher & Mountaineer, Mr. Mohammad Raashid, Bird Researcher, Mr. Shams Ul Haq Qari, Birdwatcher, Mr. Raja Aamir, Birdwatcher; Mr. Zafar Rais Mir, Researcher, Wildlife Trust of India; Mr. Mudasir Mansoor, Birdwatcher, Ms Tabish Habib, Birdwatcher, Mr. Nazir Ahmad Malik, Wildlife Guard, Dachigam National Park, and Dr. Siddharth Kaul of MoEF, New Delhi.

We also want to thank the following persons: Dr. Tej Partap, Vice Chancellor, Sher-e-Kashmir University of Agricultural Sciences & Technology of Kashmir (SKUAST-Kashmir); Dr. Shafiq A. Wani, Director Research, SKUAST; Dr. M.T. Banday, Prof. & Head, Division of LPM, SKUAST; Parvez Sajjad, Comptroller, SKUAST; Prof. G.N. Sheikh; Dr. N.A. Ganai; Mr. Parvez Ahmad Bhat, Secretary to Vice-Chancellor, SKUAST; Mr. M.R. Qadri, Assistant Executive Engineer (R&A) & I/C Automobile Workshop of SKUAST-Kashmir; Mr. Fayaz Ahmad Bhat, Assistant Comptroller/DDO, SKUAST-Kashmir, Manasbal; Mr. Sajjad Ahmad Bhat, Asst. Students Welfare Officer, SKUAST-Kashmir, Shuhama Campus; and Mr. Aflaq Hamid Wani, Assistant Professor for their cooperation.

At Wildlife Institute of India (WII), Dehradun, we would like to extend our sincere gratitude to Dr. V.B. Mathur, Director, WII; Dr. (Capt.) Parag Nigam, Scientist E & Head, Department of Wildlife Health Management; Mr. Qamar Qureshi, Scientist G/Professor, Landscape Level Planning and Management; Dr. S.P. Goyal, Scientist G/Professor and Head, Department of Animal Ecology & Conservation Biology; Dr. Bilal Habib, Scientist C, Department of Animal Ecology & Conservation Biology.

We would like to thank Mr. Masud A. Choudhary, former Vice Chancellor BGSB University; Prof. Irshad A. Hamal, Vice Chancellor BGSB University; Prof. Talat Ahmad, Vice Chancellor, University of Kashmir, and Prof. Zafar A. Reshi, Registrar, University of Kashmir.

Our gratitude to the following persons: Mr. Ravi Singh, SG & CEO, WWF-India; Dr. Sejal Worah, Programme Director, WWF-India; Dr. Parikshit Gautam, former Director Branches and Special Projects, WWF-India; Mr. A.K. Srivastava, IFS, IG, MoEF, Government of India; Dr. C.M. Seth, Chairperson, WWF-India, J&K State Office, Jammu; Ms Archana Chatterjee, National Project Coordinator, UNESCO, New Delhi; Ms Yamini Panchaksharam, Sr Programme Officer, WWF-India; Mr. Kishor Chandra, Admin. Officer, WWF-India; Ms Nisa Khatoon, Sr Project Officer, WWF-India Field Officer, Leh; Mr. Pushpinder Singh Jamwal, Project Officer, WWF-India, Field Station Gharana, Jammu; Mr. Rohit Rattan, Project Officer, WWF-India, Field Office, Pir Panjal Range, Jammu; Mr. Shakeel Ahmed, Field Assistant, WWF-India, Field Office, Pir Panjal Range, Jammu; Mr. Phuntsog Tashi, Project Officer, WWF-India Field Office, Tso Moriri, Ladakh; Mr. Tsewang Rigzin, Asst. Project Officer, WWF-India Field Office Tsokar, Ladakh; Mr. Dawa Tsering, Field Assistant, WWF-India Field Office, Tso Moriri, Ladakh; Mr. Mohd Kazim, Field Assistant, WWF-India Field Office Kargil, Ladakh; Mr. Mohd Ali, Divisional Forest Officer, Forest Department Kargil; Mr. Mohd Abbas, District Soil Conservation Officer, Forest Department Kargil; Ms Radhika Kothari, Deputy Director, Snow Leopard Conservancy India Trust; Mr. Jigmet Dadul, Programme Manager (Conservation & Livelihood), Snow Leopard Conservancy India Trust; Ms Tsering Angmo, Programme Manager (Education & Outreach), Snow Leopard

Conservancy India Trust; Ms Rigzin Chorol, General Manager (Administration & Finance), Snow Leopard Conservancy India Trust; Ms Tsering Lazes, Office Assistant, Snow Leopard Conservancy India Trust; Mr. Zafar Khan, Range Officer, Wildlife, Department of Wildlife Protection, Rajouri; Mr. Ranjeev Sharma, DFO, Wildlife Rajouri & Poonch, Department of Wildlife Protection, Rajouri; Dr. R.B. Srivastava, Scientist G, Director, Defence Institute of High Altitude Research, Leh; Dr. O.P. Chourasia, Scientist "F", Dy. Director, Defence Institute of High Altitude Research, Leh; Dr. Tsering Stobdan, Scientist 'E', Defence Institute of High Altitude Research, Leh; Mr. Anupam Anand, Research Scholar at the University of Maryland, USA; Ms Aadya Singh, Ms Vandana Thapliyal, Ms Jatinder Kaur, and Mr. Kiran Rajashekriah from Regional Programme of WWF-India; Mr. Stanzin Dorje, Director, Himalayan Film Studio, Leh; Sh. Dorjey Angchuk, Prof. Tushar Prabhu and Prof. S.P. Bagaree of the Indian Institute of Astrophysics at Hanle; Mr. S.D. Singh, Ms Kavita Suri, Ms Neelu Sharma, Mr. Shazad Khan, Mr. Vivek Gupta, Col. Vivek Chauhan, and Brig. Rajinder Singh.

We would like to offer our heartfelt thanks to our host of partners, supporters, and field workers as well as forest and wildlife officials and fellow conservationists, and above all to our IBCN partners whose support has been indispensible for the publication of this book.

Last but not least, we would like to express our gratitude to the following photographers whose images have enhanced the value of our book: Dhritiman Mukherjee, Pushpendra Singh, Ashish Kothari, Vishwatej Pawar, Saleel Tambe, K. Ramesh, Kalyan Singh Sajwan, Nikhil Devasar, Rishad Naoroji, Clement Francis, Bhasmang Mehta, Gobind Sagar Bhardwaj, Sachin Rai, Shashank Dalvi, and Chris Gomersall/RSPB.

We would like to thank BirdLife International and Royal Society for the Protection of Birds (RSPB), both based in UK, for their unstinted support to this book.

In BirdLife International, the first name that comes to mind is that of Dr. Nigel Collar whose contribution to the study of threatened birds of the world is well-known. We thank Dr. Marco Lambertini, CEO of BirdLife International, Dr. Richard Grimmett, Dr. Mike Crosby, Dr. Stuart Butcher, Dr. A.J. Stattersfield, Dr. Richard Thomas, and Mr. Joe Taylor. In RSPB, we would like to acknowledge the support of Dr. Mike Clarke, CEO, Dr. Tim Stowe, Mr. Chris Bowden, and Mr. Ian Barber.

■ ■ ■

We dedicate this book to the memory of
Dato Loke Wan Tho

born June 14, 1915, Kuala Lumpur, Malaysia
died June 1964, Taichung, Taiwan

Loke Wan Tho was a businessman, ornithologist, photographer, philanthropist and a great supporter of BNHS. He wrote many articles on birds in *JBNHS*, particularly on his visits to Jammu & Kashmir.

INTRODUCTION

India is one of the twelve mega diversity countries in the world, divided into 10 biogeographical regions based on the landmass and species distribution. The state of Jammu & Kashmir lies within the Western Himalayan and Trans-Himalayan biogeographical regions, and is placed at the junction of the temperate Palaearctic and tropical Oriental biogeographic regions of the world. More specifically, Jammu & Kashmir (32° 17' to 37° 05' N; 72° 31' to 80° 20' E) is situated in the western Himalaya, and represents the extreme west of the Himalaya in India. The state is bounded on the north by China (Karakoram Range), on the

ASAD R. RAHMANI

The scenic Kashmir Valley is one of the main tourist attractions of India. But bird tourism is still not well developed in the state

east by Tibet, to the west by Pakistan and Afghanistan, and to the south by Himachal Pradesh and Punjab. With its borders touching Pakistan, Afghanistan, Tibet, and China, the state occupies a biologically strategic spot for India. Owing to its location bordering diverse biogeographic regions of Palaearctic, Oriental, and Subtropical Punjab plains, with enormous diversity of habitat types, great mountainous heights, besides wetlands and water bodies, and diverse climatic conditions, Jammu & Kashmir is home to a vast and varied diversity of species, including some endemic and near-endemic ones.

This hilly state is divided into three geographical regions, namely, the temperate valley and mountains of Kashmir, the cold desert of Ladakh, and the

The Kashmir Valley used to have hundreds of lakes but now only the larger ones survive, such as the Wular Lake seen above

subtropical plains of Jammu. The higher regions comprise Pir Panjal, Karakoram, and the Inner Himalayan ranges. The average annual rainfall and temperature of the state has been cited as ranging from 600 to 800 mm and 15 °C to 17.5 °C respectively, but due to the varying habitats these figures are somewhat misleading. Climatic conditions vary from warm subtropical in the Jammu region to cold and arid in Ladakh. The state has a geographical area of 22.22 million ha (6.8% of India's geographical area). The total human population has increased from 10.07 million in 2001 to 12.54 million in 2010, a significant increase of about 23% in 10 years (Population Census 2010). Excluding Ladakh, which is a cold desert, the state has a forest cover of 20,230 sq. km which accounts for 20% of its total geographical area.

For administrative purposes, Jammu & Kashmir is divided into three different regions, namely Jammu, Kashmir, and Ladakh, all three of which also differ in having unique and distinctive topography, apart from their own distinctive languages, customs, and ceremonies. Thus the state of Jammu & Kashmir offers a rich diversity of landscapes, wildlife, people, and culture. There are 22 districts, 71 *tehsils*, 141 blocks, 2,690 *panchayats*, and 6,652 villages. At the local level, the *panchayats* (village councils) are the main democratically elected administrative institutions. Contrary to other states of India, Jammu & Kashmir has two capital cities. Srinagar is the summer capital, while in winter, the state machinery moves to Jammu city, the winter capital.

The state has some fabulous bird sanctuaries such as Hokarsar and Shallabugh, where it is not unusual to see four to five hundred thousand waterfowl in winter

Climatically, the state may be divided into three distinct zones: the cold arid desert of Ladakh, the temperate Kashmir Valley, and the humid subtropical region of Jammu. From the alpine (Ladakh region) to the subtropical (Jammu region), the extremes of weather and climate in Jammu & Kashmir result from its location and topographical variations.

To the south around Jammu, the climate is typically monsoonal, though the region is sufficiently far west to average 40 to 50 mm of rain per month between January and March. In the hot season, maximum temperature in Jammu city can touch 40 °C, while in July and August, very heavy though erratic rainfall occurs, with a monthly maximum up to 650 mm. In September, rainfall declines, and by October conditions are warm and extremely dry, with minimal rainfall and temperatures of around 29 °C.

From the southwest monsoon, Srinagar receives as much as 635 mm rain, the wettest months being March to May, with around 85 mm per month.

The climate of Leh, Kargil, and Zanskar districts of Ladakh is extremely dry and cold. Precipitation by rainfall is only around 100 mm per year and humidity is very low. This region, almost entirely over 3,000 msl, experiences harsh winters.

In Zanskar, the average January temperature is -20 °C, with extremes as low as -40 °C. In summer in Ladakh, daytime is typically a warm 20 °C, but with the low humidity and thin air, nights can still be cold.

Agriculture and the Economy: The state can be divided into four agro-climatic zones, namely Subtropical, Subtemperate, Temperate, and Cold Desert. Agriculture is the mainstay of the state's economy. Each region of the state has its own specific geo-climatic condition, which determines the cropping pattern and productivity. Rice is the chief crop of Kashmir and Jammu, followed by maize, barley, and wheat. The world famous Basmati rice is grown in Jammu region, which also dominates both in maize and wheat production. In the Ladakh region, barley is the major cereal crop, followed by wheat. These three important food crops, namely rice, maize, and wheat, contribute a major portion of the food grain production in the state and account for 84 percent of the total cropped area; the balance 16 per cent is shared by coarse grains and pulses. Nearly 75 per cent of the country's temperate fruits, mainly apples, are grown in the state. Other important fruits produced by the state are walnut, almond, pear, apricot, plum, cherry, peach, grapes, citrus, mango, barberry, strawberry, and quince. Some populations of chestnut, hazelnut, and cherry are found growing in their wild forms in forests. Apple however, dominates the fruit industry of the state, and at present covers an area of 127,795 hectares which constitutes 43.30% of the total area under fruit production, and 50.48 % of the total area under

ASAD R. RAHMANI

The Wildlife Department has done very good work to protect wetlands such as Gharana in Kathua distrct

Oak forests are home to many biome and restricted range species such as Kashmir Nuthatch, Spectacled Finch, and Kashmir Flycatcher

temperate fruits. This fruit accounts for 80.18% of total fruit production in the state. The area and production of fruits in the state has witnessed significant strides during the last five decades. Area under fruit crops has increased from a mere 12,400 hectares in 1953–54 to 295,141 hectares as of date. Over 500,000 families are involved in this sector. The state generates Rs. 1,700 crore income from fresh fruits and Rs. 300 crore from dry fruits, though fruit farming covers only 20 per cent of the net sown area in J&K.

Kashmir handicrafts need no introduction as these traditional crafts have made a name for themselves worldwide. This sector provides employment to about 0.2 million people. Kashmir carpets earn substantial foreign exchange (Mathew 2003). Pashmina wool, the finest natural fibre culled from the Pashmina goat, is in great demand both in national and international markets. The Changthangi and Chegu breeds of goat from Ladakh are specifically bred to produce this luxury wool. This special livestock farming serves as the main livelihood for the rural/ landless and marginal farmer community, especially the scheduled tribes of Ladakh, and the thousands of families of Kashmiri spinners, dyers, weavers, and ancillary professionals involved in the wool and shawl cottage industries.

The 300 km Srinagar-Jammu National Highway and the 265 km Mughal Road via Pir Panjal mountains are the two major surface links between the Kashmir Valley and the rest of the country. Kashmir is internationally famous for its scenic beauty, wildlife, and historic architectural monuments, and is a favourite tourist

Bird-based but strictly regulated tourism can be developed in the state with the involvement of local communities

The state has many large rivers (Jhelum, Indus, Ravi, and Chenab) and their tributaries which enrich the environment

destination, with such sites as Srinagar, Pahalgam, Gulmarg, and Sonamarg. Hindu pilgrim centres of special importance include Amarnath and Vaishno Devi. To quote Walter Lawrence (1895) "the Dal lake, measuring about 4 miles by 2½, lies close to Srinagar, and is perhaps one of the most beautiful spots in the world. The mountain ridges which are reflected in its waters, as in a mirror, are grand and varied, the trees and vegetation on the shores of the Dal being of exquisite beauty. It is difficult to say when the Dal is most beautiful" (page 21). Further on, Lawrence writes "perhaps in the whole world there is no corner as pleasant as the Dal Lake". Today, though, one needs to step out of Srinagar to enjoy such exquisite beauty.

Ladakh has also become a major tourist destination, mainly because of the unique Trans-Himalayan landscape, wildlife, and ancient and historic monasteries. Ladakh is now facing the problem of uncontrolled mushrooming of hotels, and the resultant overcrowding, litter, and degradation of its pristine habitats.

Rivers: The drainage system of the state of Jammu & Kashmir is quite significant, as is evident from the following description of its drainage pattern. The Jhelum is the main waterway of the valley of Kashmir. It has its source in a spring called Verinag, comes down the northward slopes of Pir Panjal, with a number of tributaries which make it navigable from Khanabal to Wular Lake. Its total length in the valley is 177 km, till it leaves the valley from Muzaffarabad. The Ravi is the shortest of the rivers of the state. It leaves the Himalaya at Basohli

The state has five main types of vegetation, namely Subtropical Dry Evergreen, Himalayan Moist Temperate, Himalayan Dry Temperate, Subalpine, and Alpine Forests

Agriculture and horticulture are the mainstay of the state's economy, with rice, followed by maize, barley, and wheat as the main cereal crops

in Jammu region, passing close to Kathua near Madhopur, where it enters the plains of Punjab. The Tawi river, draining the outer hill region, flows around the city of Jammu after collecting drainage to the northeast of Jammu in the interior mountains. The Chenab river originates from the Himalayan contour of Lahaul and Spiti. Two streams, more or less parallel, Chandra and Bhaga, join to form the Chandrabhaga, or Chenab. This river drains the eastern section of the southern slope of Pir Panjal. The Indus is another important river, which originates in Tibet and enters India at Demchok in Ladakh.

Vegetation: Broadly, Jammu & Kashmir has five types of vegetation, namely Subtropical Dry Evergreen, Himalayan Moist Temperate, Himalayan Dry Temperate, Subalpine, and Alpine Forests. The recorded forest area is 2.02 million ha, which constitutes 9% of the geographical area of the state. Forests are largely distributed in the Kashmir Valley and Jammu region. Dense forest and open forest account for 1,184,800 ha and 938,900 ha respectively, according to the Ministry of Environment and Forests (2001). There are 22 districts in the state. The western districts have more forest cover with dense and open forests. Reasi, Poonch, Kathua, and Jammu have more forest cover than Ladakh, Gilgit, Baramulla, and Anantnag (Ministry of Environment and Forests 2001). Ladakh region, being part of Trans-Himalaya, is devoid of forest vegetation, although this area is now being extensively planted with Willow *Salix* spp., which is affecting avifauna by changing the natural habitat. Ladakh region is known to be a cold desert with highly adapted flora and fauna (Gujja *et al.* 2003). The Mediterranean mixed coniferous forests and high altitude meadows are bestowed with a unique wealth of plant species of medicinal and aromatic value, like the Yew *Taxus baccata*, Veer Tethwan or Krimidru *Artemisia amygdalina*, *Artemisia indica*, *Artemisia* spp., Mexican Tea or Jangli Javind *Chenopodium ambrosioides*, Brahmakamal, Bonnet Bellflower, or Branmool *Codonopsis ovata*, Laljari, Hound's Tongue, or Neelkhand *Cynoglossum glochidiatum*, Ratanjot or Gaozaban *Arnebia benthamii*, Ratanjot or Kahzaban *Onosma hispidum*, Suranjan or Virkimposh *Colchicum luteum*, and Costus root or Kuth *Saussurea costus*.

The most magnificent tree of Kashmir is the Chinar, introduced during the Mughal period and now found throughout the valley, which grows to gigantic size and girth. Walnut, Willow, Almond, Poplar, Pine, and Cedar also add to the rich flora of Kashmir. A valuable tree is the Willow (used to make cricket bats and wickerwork), which mostly grows in marshy areas along rivers and streams. Willow is also found on river banks in Leh and Kargil districts of Ladakh region, and in Poonch, Doda, and Kishtwar districts of Jammu region.

Wetlands and Waterbodies: The Greater Himalaya and Trans-Himalaya (1,500 to 6,000 msl) consisting of Kashmir Valley and Ladakh high altitude region is a source of many major rivers, such as the Jhelum and the Indus, which originate from this region, but much of the water is internally drained where the rivers end in vast lakes and marshes.

Inland wetlands of Jammu & Kashmir cover an area of 406,780 ha and include among others the 11 wetland reserves fulfilling Ramsar criteria: Chushul Marshes,

Brackish wetlands of Ladakh region provide nesting habitats to Black-necked Crane, Bar-headed Goose, and many other species

Environmental education should be started from school level. Presently it is undeveloped in Jammu & Kashmir

Haigam, Hanle, Hokarsar, Mirgund, Pangong Tso, Shallabugh, Surinsar-Mansar, Tso Kar, Tsomoriri Lake, and Wular Lake (Islam & Rahmani 2008). Four of them are already Ramsar Sites. These wetlands, besides supporting many species of wetland plants, animals, and insects, provide critical habitats for the life cycle of ducks, geese, swans, and many other waterfowl species. Such lakes and marshes, mostly freshwater and saline respectively, are important as wintering and breeding grounds for a diversity of waterfowl such as the endangered Black-necked Crane *Grus nigricollis*, Bar-headed Goose *Anser indicus*, and Great Crested Grebe *Podiceps cristatus*.

The Tibetan Sandgrouse *Syrrhaptes tibetanus* and Tibetan Partridge *Perdix hodgsoniae*, representing Biome 5, occur on the surrounding dry plains.

The wetlands of Kashmir Valley are important for both resident and migratory waterfowl. They are major wintering areas for a variety of migratory ducks and geese, and extremely important breeding areas for Mallard *Anas platyrhynchos*, Blunt-winged Warbler *Acrocephalus concinens* (Holmes & Parr 1988) and Ferruginous Duck *Aythya nyroca* (Bates & Lowther 1952), besides a variety of other waterfowl. They are particularly important for long distance migrants, as a stopover site for feeding and resting. Many waterbirds occur in huge numbers in the wetlands of the state, much above the proportion of the total species abundance (1%) determined by Wetlands International (2006) as one of the criteria for declaring a wetland as a Ramsar Site.

In Jammu region, Gharana, Mansar, Surinsar, Nandansar, Chandansar, Kotarisar, besides supporting many species of wetland plants and animals, provide critical habitats for ducks, geese, swans, and many other waterfowl species.

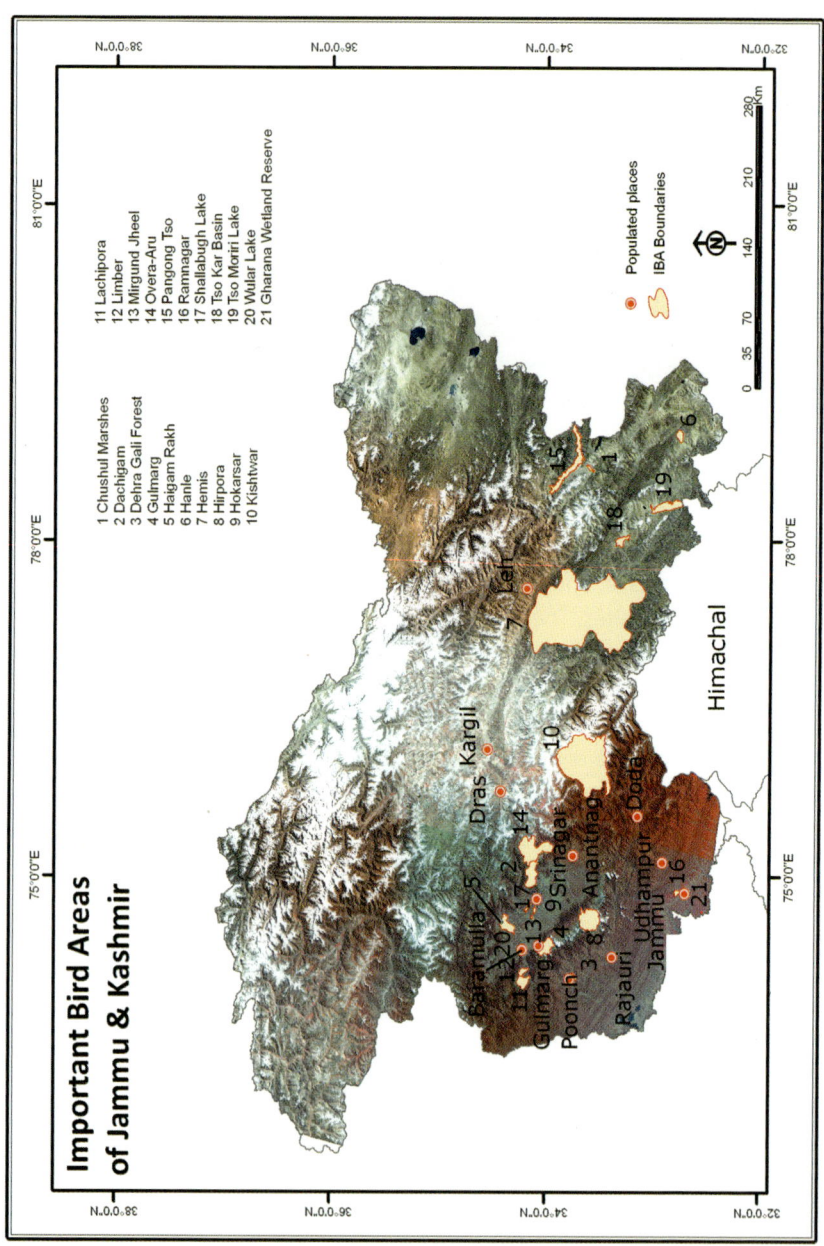

Important Bird Areas
of Jammu & Kashmir

1 Chushul Marshes
2 Dachigam
3 Dehra Gali Forest
4 Gulmarg
5 Hajgam Rakh
6 Hanle
7 Hemis
8 Hirpora
9 Hokarsar
10 Kishtwar

11 Lachipora
12 Limber
13 Mirgund Jheel
14 Overa-Aru
15 Pangong Tso
16 Rannagar
17 Shallabugh Lake
18 Tso Kar Basin
19 Tso Moriri Lake
20 Wular Lake
21 Gharana Wetland Reserve

Populated places
IBA Boundaries

0   35   70   140   210   280
                                          Km

Himachal

Threatened Birds of Jammu & Kashmir and their presence in IBAs

Critically Endangered	IBA code
White-backed Vulture *Gyps bengalensis*	IN-JK-16
Slender-billed Vulture *Gyps tenuirostris*	IN-JK-16

Endangered

Egyptian Vulture *Neophron percnopterus*	IN-JK-02, 08
White-headed Duck	IN-JK-20

Vulnerable

Marbled Teal *Marmaronetta angustirostris*	IN-JK-20
Pallas's Fish-eagle *Haliaeetus leucoryphus*	IN-JK-05, 09, 20
Greater Spotted Eagle *Aquila clanga*	IN-JK-06
Eastern Imperial Eagle *Aquila heliaca*	IN-JK-02
Western Tragopan *Tragopan melanocephalus*	IN-JK-03, 10, 11, 12
Cheer Pheasant *Catreus wallichii*	IN-JK-12
Sarus Crane *Grus antigone*	IN-JK-21 (?)
Black-necked Crane *Grus nigricollis*	IN-JK-01, 06, 15, 18, 19
Kashmir Flycatcher *Ficedula subrubra*	IN-JK-02, 03, 04, 08, 14

Near Threatened

Oriental Darter *Anhinga melanogaster*	IN-JK-20
Ferruginous Pochard *Aythya nyroca*	IN-JK-05, 19, 20, 09
Black-bellied Tern *Sterna acuticauda*	IN-JK-05, 09, 13, 17, 20
European Roller *Coracias garrulus*	IN-02, 14, 04, 08, 11, 12
Tytler's Leaf-warbler *Phylloscopus tytleri*	IN-02, 14
Long-billed Bush-warbler *Bradypterus major*	IN-JK-02 (unconfirmed report)

Avifauna of Endemic Bird Area 128: Western Himalaya in IBAs of J&K

Western Tragopan	*Tragopan melanocephalus*	IN-JK-03, 10, 11,12
Cheer Pheasant	*Catreus wallichii*	IN-JK-12, 10
Tytler's Leaf-warbler	*Phylloscopus tytleri*	IN-JK-02, 14
Kashmir Flycatcher	*Ficedula subrubra*	IN-JK-02, 03, 04, 08, 14
Kashmir Nuthatch	*Sitta cashmirensis*	IN-JK-14, 02
Orange Bullfinch	*Pyrrhula aurantiaca*	IN-JK-02, 14

According to estimates by the Ministry of Environment and Forests, the state of Jammu & Kashmir has 29,107 ha area under wetlands, out of which 7,227 ha is under natural and 21,880 ha under man-made wetlands. The biological and

Globally Vulnerable Pallas's Fish-eagle has almost disappeared from the state, with very few recent records

socio-economic importance of the wetlands of Jammu & Kashmir necessitate the identification and prioritisation of some representative wetlands which urgently need conservation and carefully planned use. In 1998, the Space Applications Centre (ISRO) Ahmedabad, using remote sensing technology, identified and mapped 42 wetlands in J&K which were greater than 56.25 ha, occupying a total area of 406,779.36 ha. In addition, 38 wetlands smaller than the minimum mapping unit, i.e. 56.25 ha, were detected and indicated on the maps. The Directorate of Environment and Remote Sensing, Government of Jammu & Kashmir, using Survey of India maps of 1976 have listed about 1,248 wetlands of varying sizes.

IBAs and Protected Areas: Jammu & Kashmir has been a pioneering state in the field of conservation, with a network of wildlife protected areas (the erstwhile game reserves) established during the reign of the Maharajas long before Independence. These game reserves were covered by the Game Preservation Act, 1852 which was revised and updated as the J&K Wildlife Protection Act, 1978 (Amended 2002). Some of the sanctuaries were established nearly one hundred years ago, mainly to protect the catchment areas of important lakes and to provide hunting grounds for the Maharajas. Since Independence, the state Government has notified about 16,000 sq. km under the Protected Area Network (PAN), which is 15.6% of the total geographical area of the state, comprising four National Parks, 14 Wildlife Sanctuaries, and 35 Conservation Reserves. The Protected Areas (PAs) include 2,762 sq. km of forest (12.76% of 20,230 sq. km total forest area). The remaining 13,150 sq. km is the high altitude cold desert area of Ladakh. Jammu & Kashmir may be credited with having the largest number of PAs (53) in the country,

As in other parts of India, the White-backed Vulture has declined drastically in Jammu & Kashmir

DHRITIMAN MUKHERJEE

with the 15.6% coverage of PAs in J&K being nearly three times more than the national average. In terms of the geographical area covered by the PAs, the state is second only to Gujarat (J&K Wildlife Protection Department 2010 unpublished report).

Of these Protected Areas, Dachigam National Park is of special ecological significance as it harbours the last viable population of the globally Threatened Kashmir Red Deer or Hangul *Cervus elaphus hanglu*. Wular Lake, situated in Baramulla district, comprising an area of 8,900 ha, is a wetland of international importance which was declared as one of the first six Ramsar Sites in India in 1990. Of the four national parks, three have been identified as IBAs. Of the 16 wildlife sanctuaries, eight are IBAs. Of the 21 Important Bird Areas (IBAs) identified in 2004 (Islam & Rahmani 2004) and seven potential IBA identified in 2012 (Rahmani *et al.* 2012), 11 fulfill Ramsar criteria (Islam & Rahmani 2008).

AVIFAUNA

Jammu & Kashmir, besides harbouring some species that are shared with the tropical and subtropical parts of India, is home to an equally impressive faunal diversity, particularly among birds and mammals unique to higher altitudes. Some of these mammals and birds are endemic to relatively limited areas of Western Himalaya.

Jammu & Kashmir lies in the Western Himalaya Endemic Bird Area (EBA 128) where 11 Restricted Range species have been listed by Stattersfield *et al.* (1998). Because of great altitudinal variation and differing physiogeographical regions,

Many large, well-drained valleys of Ladakh are under human occupation. As Buddhist culture encourages people to live in harmony with nature, birds survive in these valleys

Jammu & Kashmir has three biomes: Biome 5 (Eurasian High Montane – Alpine and Tibetan) above 3,600 msl; Biome 7 (Sino-Himalayan Temperate Forest) mainly *c.* 1,800 to 3,600 msl; and Biome 8 (Sino-Himalayan Subtropical Forest) between *c.* 1,000 and 2,000 msl. The Eurasian High Montane (Alpine and Tibetan) Biome is mainly distributed in the Ladakh region, especially in Changthang plateau. Sino-Himalayan Temperate Forest habitat is present in most of the IBAs in the State.

Kashmir Valley: In the Kashmir Valley, many protected areas support Restricted Range species and some waterbodies support large congregations of migratory waterbirds. These Restricted Range species occur mainly in Temperate Coniferous or Broadleaf Forest, Subalpine Forest, and Montane Grasslands. For example, the Kashmir Flycatcher *Ficedula subrubra,* which is considered a globally Threatened species, is found between 1,800 and 2,700 msl in Temperate Mixed Broadleaf Forest, especially where there is dense growth of Parrotia *Parrotiopsis jacquemontiana* (Stattersfield *et al.* 1998). Other similar species, namely Tytler's Leaf-warbler *Phylloscopus tytleri* and White-throated Tit *Aegithalos niveogularis* are found between 1,500 and 3,600 msl in Pine-Oak Mixed Deciduous Forests. Other Restricted Range species which are found in or near the Valley are the Kashmir Nuthatch *Sitta cashmirensis*, Spectacled Finch *Callacanthis burtoni*, and Orange Bullfinch *Pyrrhula aurantiaca*. The two finches are found in open Coniferous Forest, Mixed Woodland Forest, Deciduous Forest, and occasionally in stands of Birch *Betula utilis* (Stattersfield *et al.* 1998).

Flat gravel ground with low vegetation cover is an ideal habitat for Restricted Range species such as Tibetan Sandgrouse

Ladakh: The Changthang region in Ladakh is an important breeding ground for waterbirds. Apart from hosting the largest breeding congregation of Bar-headed Geese *Anser indicus* in India, Changthang also supports the largest population of the globally Vulnerable Black-necked Crane *Grus nigricollis* in India (Chandan *et al.* 2008). During a study on its breeding ecology, Pfister (1998) recorded 12 sites in the Changthang region as breeding sites of this species and counted 38 individuals of Black-necked Crane. In a subsequent survey of Changthang in 2001, 42 individuals were counted, with 10 breeding pairs in the Changthang region (S.A. Hussain *pers. comm.* 2003). During subsequent years, a study on the status and ecology of the Black-necked Crane was conducted by WWF-India. During the 2012 WWF-I survey, a population of 139 Black-necked Crane was recorded in Ladakh.

Changthang is a huge area and many wetlands and other important spots (e.g., Sumdo near Puga) are included in this IBA. There are many small wetland sites which are important breeding grounds for waterbirds but do not fulfill IBA criteria, so the Changthang plateau as a whole could be considered one IBA. This does not mean that the individual wetlands are not important. All the sites are part of the IBA and should be well protected. Otto Pfister has suggested that the Changthang area should be divided into three regions, from the conservation point of view. These are: i. Pangong Tso region (northern Changthang Wilderness

Area, including Pangong Tso, Chushul, Harong, and Lungparma), Hanle region (eastern Changthang Wilderness Area, including Hanle plain, Lalpari, Staklung, and Fukche), and Tso Moriri region (western Changthang Wilderness Area, including Tso Kar plain, Puga, Tso Moriri, and Chumur).

Hemis National Park in Ladakh is another vital IBA site, which is important for all the high altitude birds of the Western Himalaya. About 80 bird species are found in the Park, and 50 of them breed there. The important bird species recorded in 1999 by Khursheed Ahmad (*pers. comm.* 2012) in Markha Valley within Hemis National Park include Bearded Vulture or Lammergeier *Gypaetus barbatus*, Red-billed Chough *Pyrrhocorax pyrrhocorax*, Yellow-billed Chough *Pyrrhocorax graculus*, Robin Accentor *Prunella rubeculoides,* Common Great Rosefinch *Carpodacus rubicilla*, Streaked Great Rosefinch *C. rubicilloides*, Black-throated Thrush *Turdus ruficollis*, and Brown Accentor *Prunella fulvescens*.

Jammu region: It is characterised by mixed types of forest. The districts of Udhampur, Jammu, Kathua, Reasi, and Poonch are very important for Biome 8 species. Detailed studies have not been conducted, and very little is known about the birds here, except for a couple of rapid surveys by M.M. Baba (Wildlife Warden) in 1999–2000. The area between 500 and 1,500 msl is particularly denuded, and certainly requires more study (Trevor Price *pers. comm.* 2011). Ramnagar Wildlife Sanctuary is one of the most important IBAs in Jammu region, and is a lifeline for the people of Jammu. It harbours 30 species of breeding birds, besides a variety of other important biodiversity.

Potential IBA Areas

Recently, Rahmani *et al.* (2012) identified the following seven sites as potential IBAs:
1. Aharbal-Kounsarnag Forests
2. Gurez Valley
3. Kanji Wildlife Sanctuary
4. Rangdum Wetlands
5. Sheshera Reserve Forest
6. Shikargah Conservation Reserve, Tral
7. Suru Valley

Threatened Species

Historically, in Jammu & Kashmir, 18 globally Threatened species have been recorded (updated to 2011), such as the Siberian Crane *Grus leucogeranus* from Leh, White-headed Duck *Oxyura leucocephala* and Lesser White-fronted Goose *Anser erythropus* from Wular Lake, Baikal Teal *Anas formosa* from Mirgund Jheel, Marbled Teal *Marmaronetta angustirostris* from Wular Lake and Mirgund Jheel, and Pallas's Fish-eagle *Haliaeetus leucoryphus* from Wular Lake, Haigam Rakh,

Leh, Hokarsar, Chushul, Marbul Pass, Tso Moriri Lake, and Hanle. The Greater Spotted Eagle *Aquila clanga* was also reported from Bhaderwah in Jammu (BirdLife International 2001) and the Eastern Imperial Eagle *Aquila heliaca* from Kashmir Valley. Cheer Pheasant (*Catreus wallichii*) has been reported from Limber Wildlife Sanctuary and Kishtwar National Park. Sarus Crane *Grus antigone* was reported from Kathua district, at Kishanpur Garuna Wetland Reserve (Choudhury *et al.* 1999, Sahi 1993, Sundar 1999); Eastern Stock Pigeon *Columba eversmanni* was reported from Limber Wildlife Sanctuary (Javed 1992). Some of these species are still reported from J&K, but birds such as the Siberian Crane, White-headed Duck, Lesser White-fronted Goose, and the Marbled and Baikal Teal have not been reported for quite some time.

In recent years, 12 globally Threatened and six Near Threatened species have been recorded from the IBAs in Jammu & Kashmir. Most of the species are widespread, such as Pallas's Fish-eagle, Eastern Imperial Eagle, White-backed Vulture, and Greater Spotted Eagle. Black-necked Crane is found in the Changthang plateau in small numbers, their main population being in Tibet. Along with neighbouring Himachal Pradesh, Jammu & Kashmir is extremely important for the long-term survival of Western Tragopan *Tragopan melanocephalus* and Cheer Pheasant *Catreus wallichii*. Kashmir Flycatcher *Ficedula subrubra* is another vital element of the avifauna, its main breeding population being recorded in J&K, in Overa Wildlife Sanctuary (Price & Jamdar 1990), and in Dachigam National Park in 2003 (Khursheed Ahmad *unpubl.*, Trevor Price *pers. comm.* 2012).

Restricted Range species: In the Western Himalaya (Endemic Bird Area 128), the main habitats are the Temperate Coniferous or Broadleaf Forest, Subalpine Forest, and Montane Grassland. These habitats have 11 Restricted Range avian species between 1,500 and 3,600 msl. Of these, four are globally Threatened (Stattersfield *et al.* 1998, BirdLife International 2001). In Jammu & Kashmir, 10 Restricted Range species are found, the exception being Himalayan Quail *Ophrysia superciliosa* (thought to be extinct), which used to be distributed in long grass and brushwood on steep hillsides, and was reported from Uttarakhand over a century ago. Other Restricted Range species in Jammu & Kashmir are the Western Tragopan, which inhabits dense undergrowth in Coniferous, Mixed, and Oak Forests between 1,350 msl (in winter) and 3,600 msl; Cheer Pheasant on steep grassy slopes, Open Coniferous or Deciduous Forests, is habituated to early successional habitats from 1,400 to 3,500 msl; Brooks's Leaf-warbler *Phylloscopus subviridis* has not been reported from any of the IBAs from its habitat of Coniferous and Mixed Forest in drier, cooler areas between 2,100 and 3,600 msl; Tytler's Leaf-warbler *Phylloscopus tytleri* is reported from Dachigam National Park in Coniferous Forest, with dwarf willows and birches near the tree line from 2,400 to 3,100 msl. The breeding of Kashmir Flycatcher *Ficedula subrubra* is restricted to Kashmir, but it winters as far south in India as Tamil Nadu and Kerala, and extralimitally in Sri Lanka. In Jammu & Kashmir, it is reported from Dachigam

National Park (Gauntlett 1972, Ahmad 1999), Dehra Gali in Jammu (Tahir Shawl *pers. comm.* 2003), Gulmarg Wildlife Sanctuary, Overa-Aru Wildlife Sanctuary, and Hirpora (Tahir Shawl *pers. comm.* 2003) in Temperate Mixed Broadleaf Forest, especially where there is dense growth of Parrotia between 1,800 and 2,700 msl. White-cheeked Tit *Aegithalos leucogenys* is not reported from any IBA, but Spectacled Finch *Callacanthis burtoni* is regularly seen in Overa (Price & Jamdar 1990, Price *et al.* 2003). The White-throated Tit *Aegithalos niveogularis* has been regularly seen in Overa Wildlife Sanctuary near the tree line (Price *et al.* 2003). Orange Bullfinch *Pyrrhula aurantiaca* and Kashmir Nuthatch *Sitta cashmirensis* have been recorded from Dachigam National Park (Khursheed Ahmad *unpubl.* 2012). The Orange Bullfinch is also regularly seen Overa-Aru Wildlife Sanctuary in Open Coniferous and Mixed Forest. Kashmir Nuthatch, though earlier reported only from Overa-Aru (Price & Jamdar 1990), has also been sighted in Dachigam in the subalpine meadow at *c.* 3300 msl (Ahmad 1999, Khursheed Ahmad *unpubl.* 2003).

Biome 5

Ladakh lies in Biome 5 (Eurasian High Montane – Alpine and Tibetan) where BirdLife International (undated) has identified 48 species that represent the bird assemblages of this biome. Pfister (2004) published a checklist of 25 species of Biome 5 in Ladakh. In and around the Changthang area, thick stands of *Hippophae rhamnoides* and other vegetation in the Markha and Chang Chu Valleys are present, which have important habitats for large numbers of wintering passerines. These include Guldenstadt's Redstart *Phoenicurus erythrogaster*, Common Great Rosefinch *Carpodacus rubicilla*, Streaked Great Rosefinch *C. rubicilloides*, Black-throated Thrush *Turdus ruficollis*, Stoliczka's Tit-warbler *Leptopoecile sophiae*, Robin Accentor *Prunella rubeculoides*, and Brown Accentor *Prunella fulvescens* (Mallon 1987, 1989).

THREATS AND CONSERVATION ISSUES

As in other parts of India, drastic increase in the human and livestock population in J&K has created pressures on natural resources, and thereby forests, pastures, and grasslands have been brought under cultivation to sustain the increased demand for cereals and other food products. Due to the biotic interferences from unsustainable land use patterns in rural areas and encroachment upon wild lands, agriculture, excessive livestock grazing, and other extractive industries, most protected areas became fragmented, degraded, and disturbed, and these habitat modifications have caused the ecological dislocation of many wildlife species. These habitat modifications have far reaching and negative impacts on wildlife, and have caused many species to become locally extinct.

With animal husbandry providing livelihood to numerous families, over-grazing is a huge problem in the whole state

Among the several threatened species inhabiting this region is the Hangul, which is endemic to Kashmir. Musk Deer and Markhor, Cheer Pheasant, Western Tragopan, and Black-necked Crane are other Threatened species that require immediate management and conservation inputs.

In Jammu & Kashmir, wildlife conservation in general and Threatened species (particularly Hangul) conservation has been given priority in all management plans, beginning with the first one drafted in 1971 by Collin Holloway. The population of Hangul in Kashmir has been reduced from an estimated 2,000 individuals in 1947 to about 140 to 170 individuals at present (Ahmad *et al.* 2009). Livestock grazing is a major problem in all the protected areas and IBAs in J&K. Even in the prestigious Dachigam NP, there is a sheep farm within the notified area. Despite repeated attempts, the Wildlife Department has not succeeded in relocating it outside the national park. Other problems include the lack of coordination between the numerous departments that are stakeholders, logically or otherwise, in the Park (Animal Husbandry, Hospitality and Protocol, PWD, Irrigation and Water Works, Electricity, Telephones, Agriculture, and Fisheries). Major disturbance is caused to wildlife by visitors driving noisily along the 5 km stretch of road to the VIP lodge at Draphama (Gruisen 1983). At present, the army and paramilitary forces have a base inside the national park, and they not only occupy the accommodation meant for frontline wildlife staff, but also cause disturbance to the Hangul habitats, particularly during the breeding season. This

Cutting of hillsides for road-building plays havoc with the ecology. Such destruction can be easily avoided by strict compliance to environmental laws

situation prevails in other protected areas and IBAs also.

In Ladakh, unplanned developmental activities are the main concern in the Changthang region. Plantation in the marshes for the wickerwork industries, and construction of roads directly affects the breeding grounds of the Vulnerable Black-necked Crane. Increasing settlements near the crane's nesting habitats have resulted in an increase in the feral dog population, a major predator on crane eggs and chicks in Changthang.

Overall, the key threats to birds and other biodiversity of the state are habitat encroachment, overgrazing by livestock, tourism, firewood collection, and forest fires. As is the case across the world and in India, loss of wetlands owing to human interference is continuing and has resulted in the degradation of large numbers of wetlands in the state. Besides numerous human disturbances threatening the survival of wetlands in Kashmir, such as increased siltation, eutrophication, overfishing, and the encroachment of agricultural land into the marshes peripheral to lakes, climate change is reported to have impacted wetlands in Jammu & Kashmir, as also worldwide, by drying up small wetlands resulting in the loss of these "carbon sinks"; converting permanent wetlands into seasonal wetlands, subject to greater variation in water levels; enhancing the release of greenhouse gases from these systems; and decreasing biodiversity within affected wetlands.

GENERAL RECOMMENDATIONS FOR BIRD SPECIES/HABITAT CONSERVATION IN JAMMU & KASHMIR

In order to promote conservation of birds in general and threatened birds in particular in the state of Jammu & Kashmir, the following steps are recommended:

1. Strict control on hunting and poaching through enforcement of available legislative framework.

2. Inventorisation of the avifauna: proper surveys of different habitats across latitudinal and altitudinal gradients should be conducted in order to compile an authentic inventory of the avifauna of J&K.

3. Research on avifauna must focus on:

    a. Baseline data to ascertain the status of the avifauna of J&K.

    b. Long-term and periodic monitoring of the status of the birds, especially threatened species.

    c. Ascertain level of threats to different threatened birds of J&K.

    d. Mapping of critical bird habitats using Remote Sensing and GIS for supporting decision making and effective management.

    e. In case of bird species with unknown migratory and local movement patterns, we must employ ringing/tagging and satellite telemetry techniques to track their dispersal and movements.

    f. Improved understanding of the habitat requirements and protection of breeding habitat in case of species for which habitat decline is a threat.

4. Species-specific action plans: In order to make conservation efforts judicious, effective, and successful, it is mandatory to involve subject experts in preparing species-specific action plans to guide enforcement agencies regarding workable conservation strategies.

The Wildlife Department of J&K promotes state-of-the art technology in wildlife research. Mr. Intesar Suhail, Wildlife Warden of Leh (left) releasing Bar-headed Goose with neck collar and satellite tag

Wetlands are unfortunately still considered 'wasteland' to be used for throwing garbage

5. Since birds as a group are most visible and ideal indicator species, the state must consider sponsoring long-term research projects to investigate the effects of climate change on the bird communities, bird dispersal, movement patterns and ecological behaviour.

6. In view of the general apathy of the state, local populace, and policy makers, there is a need to raise public awareness through a variety of tools such as lectures, advertisements, bird fairs, workshops, printed literature distribution, popular birdwatching events, seminars, and orientation courses for school and university teachers and policy makers.

7. Promote conservation action through the involvement of stakeholder communities in and around the protected area network as well as in the non-protected IBAs.

8. Control insensitive tourism related activities in important bird habitats through law enforcement. Also train and involve local communities in promoting ecotourism. While a beginning in this direction has already been made in the state, effective implementation of the policies on the ground is awaited in order to effectively shield threatened bird species and their vulnerable habitats from unregulated tourism.

9. Habitat improvement wherever habitat loss or degradation is contributing towards the decline of the species should be a priority.

10. The Department of Wildlife Protection, Government of J&K, must organise programmes for training and capacity building of the field staff, in order to enable them in bird identification and bird monitoring techniques. Since the majority of the field staff is transferred from other wings of the Forest Department, successful

Threatened Birds of Jammu & Kashmir

completion of training and capacity building programmes should be compulsory criteria for placement in the Wildlife Department.

11. The Department of Wildlife Protection must consider organising Orientation Courses for new entrants and Refresher Courses for in-service staff in order to update their knowledge about bird diversity of the state, conservation, and monitoring techniques.

12. In order to ensure that the Critically Endangered Gyps vultures are not wiped out from the state, it would be important to strictly enforce ban on the use of diclofenac for veterinary use, and promote the use of the safe alternative meloxicam.

13. In view of the well documented ill effects of the rampant use of pesticides and inorganic fertilisers on ecosystems, it is immensely important for the state to encourage organic farming around wetlands specifically in Kashmir Valley as well as in other regions.

14. Conservation and restoration of wetlands should be a priority in order to maintain ecological services and conserve the biological diversity that is directly and indirectly dependent on such ecosystems.

15. Reduce anthropogenic pressure in important birdlife habitats, especially during the breeding seasons, in order to minimise potential threats to breeding birds.

16. Avoid unplanned development activities in and around the areas identified as Important Bird Areas (IBAs) and Protected Areas (PAs). Unregulated and unwanted development can open up and expose the areas and species to unprecedented threats.

17. Ecologically important wetland habitats of Jammu & Kashmir state have been facing invasion by alien invasive plants. Needless to mention that very little has been done by the state to eradicate or contain the biological invasion of its pristine wildlife habitats. Some efforts have been made in this regard in some wetlands without any significant results. Therefore this area deserves special attention from the state authorities.

ASAD R. RAHMANI

Many wetlands of the Kashmir Valley, such as Wular above, have been planted over with fast growing trees by the Forest Department. This practice should be stopped immediately

Unregulated tourism with off-the-road driving is destroying the fragile ecosystem of Ladakh. Sometimes Marmots (inset) are run over by careless drivers

18. In case of bird species and their habitats located along the geographical borders of the state, it is important to involve the armed forces, specially the ITBP and Indian Army, in conservation efforts. This can be achieved by undertaking collaborative projects and fostering concern among the armed forces through awareness generation and capacity building.

19. Bird Literature: Development of literature on birds and other wildlife in Urdu, Hindi, Kashmiri, and Ladakhi. There is no fieldguide to the birds of Jammu & Kashmir. Similarly, bird books for children can be brought out in local languages to create interest in birds from the beginning. Urdu and Kashmiri literature and poetry are full of mention of birds, their beauty and melody.

Due to its fragile ecosystem, there should be strict dermarcation of camping sites in Ladakh. All garbage should brought back to Leh for proper disposal

PUSHPINDER SINGH JAMWAL

PANKAJ CHANDAN

Stray dogs have become a big problem all over the state, particularly in Ladakh where nearly 50% of the chicks of Black-necked Crane are killed by them

20. Development of bird tourism policy: Tourism is the back-bone of the economy of the state. Bird tourism is a multi-billion industry in the world with millions of people travelling for birdwatching to different destinations. Specialised tour operators in India are catching up to capture this niche market. Bird tourism also helps local communities by providing them direct employment (e.g., tourist guide) or in hospitality industry (e.g., homestay concept in Ladakh).

21. Monitoring of pesticide levels in common birds: Huge amounts of pesticides are used in the fruit industry of the state, particularly for apple and apricot, but their impact on birds is not known. Similarly, pesticides are used in paddy cultivation. A long-term monitoring of pesticide levels in birds is needed to know negative impacts, if any.

22. Surveillance of avian flu and other avian diseases by regular sampling of migratory birds: With new emerging avian diseases and recurrence of H5N1 on regular basis, it is imperative that avian disease surveillance should become a regular programme of the Wildlife Department in collaboration with the BNHS, WWF, and veterinarians. Blood samples thus collected should be sent to High Security Animal Disease Laboratory, Bhopal for analysis according to their protocol.

23. Ringing/Marking Programme: Long-term bird ringing/marking programme in Gharana and other wetlands, and also in the wetlands of Ladakh should be implemented. Regular bird ringing with the collaboration of BNHS will help to track the movement of migratory birds, and to collect blood samples for avian disease surveillance. Activities like bird banding generate public interest in bird conservation and also inculcate scientific attitudes towards data collection.

24. University Course: Development of curriculum on wildlife at BSc and MSc levels have been started in many universities/states but there is no such course in Jammu & Kashmir to train the younger generation.

Zanskar mountain range spreads about 7,000 sq. km and ranges from 3,500 to 7,000 m

ASHISH KOTHARI

Although it is not yet included in the Red List by IUCN, Ibisbill appears to have declined in Ladakh due to human disturbance

25. Conservation Education Centres: Development of nature interpretation and conservation education centres near all major cities and protected areas, particularly in Srinagar and Jammu. These centres should be designed, developed, and operated on public-private partnership basis. They should be made self-sufficient financially by charging appropriate fees for school and college visits.

ASAD R. RAHMANI

Well-intentioned but totally unnecessary conservation actions like creating fenced zones to protect nests of birds should be avoided

REFERENCES

Ahmad, K. (1999) Birds of Dachigam National Park. *Newsletter for Birdwatchers* 39(2): 22–24.

Ahmad, K., Sathyakumar, S. and Qureshi, Q. (2009) Conservation status of the last surviving population of Hangul (*Cervus elaphus hanglu*) in Kashmir. *J. Bombay Nat. Hist. Soc.* 106(3): 245–255.

Bates, R.S.P. and Lowther, E.H.N. (1952) *Breeding birds of Kashmir*. Oxford University Press, Oxford.

BirdLife International (2001) *Threatened Birds of Asia: The BirdLife International Red Data Book*. 2 vols. BirdLife International, Cambridge, UK.

BirdLife International (undated) *Important Bird Areas (IBAs) in Asia: Project briefing book*. BirdLife International, Cambridge, UK. Unpubl.

Chandan, P., Chatterjee, A. and Gautam, P. (2008) Management planning of Himalayan high altitude wetlands. A case study of Tsomoriri and Tsokar wetlands in Ladakh, India. *Proceedings of Taal 2007: The 12th World Lake Conference*: 1446–1452.

Choudhury, B.C., Kaur, J. and Gopi Sundar, K.S. (1999) Sarus Crane Count-1999. Wildlife Institute of India, Dehradun.

Gauntlett, F.M. (1972) Notes on some Kashmir birds. *J. Bombay Nat. Hist. Soc.* 69: 591–615.

Gujja, B., Chatterjee, A., Gautam, P. and Chandan, P. (2003) Wetlands and lakes at the top of the world. *Mountain Research and Development* 23(3): 219–221.

Gruisen, J. van (1983) The Hangul, Dachigam's endangered deer. *Sanctuary (Asia)* 3: 114–31.

Holmes, P.R. and Parr, A.J. (1988) A checklist of the birds of Haigam Rakh, Kashmir. *J. Bombay Nat. Hist. Soc.* 85: 465–473.

Islam, M.Z. & Rahmani, A.R. (2004) *Important Bird Areas in India: Priority sites for conservation*. IBCN, BNHS, BirdLife International and Oxford University Press, Mumbai. Pp. xvii + 1133.

Islam, M.Z. & Rahmani, A.R. (2008) *Existing and Potential Ramsar Sites of India*. Indian Bird Conservation Network, Bombay Natural History Society, BirdLife International and Royal Society for the Protection of Birds. Oxford University Press. Pp. 592.

Javed, S. (1992) Birds of Limber valley forest (Jammu and Kashmir). *Newsletter for Birdwatchers* 32(5/6): 13–15.

Lawrence, W.R. (1895) *The Valley of Kashmir*. H. Frowde, London.

Mallon, D.P. (1987) The winter birds of Ladakh. *Forktail* 3: 27–41.

Mallon, D.P. (1989) An ecological survey of the protected area network in Ladakh. Report to the Department of Wildlife Protection, Jammu & Kashmir. Unpubl.

Mathew, K.M. (ed.) (2003) Manorama Yearbook 2003. Malayale Manorama, Kottayam.

Ministry of Environment and Forests (2001) *State of Forests*. Forest Survey of India, Dehra Dun.

Pfister, O. (1998) The breeding ecology and conservation of the Black-necked Crane (*Grus nigricollis*) in Ladakh, India. Unpubl. Thesis. University of Hull, Hull, UK. Pp. 136.

Pfister, O. (2004) *Birds and Mammals of Ladakh*. Oxford University Press, New Delhi.

Price, T.D. and Jamdar, N. (1990) The breeding birds of Overa Wildlife Sanctuary, Kashmir. *J. Bombay Nat. Hist. Soc.* 87: 1–15.

Price, T., Zee, J., Jamdar, K. and Jamdar, N. (2003) Bird species diversity along the Himalaya: A comparison of Himachal Pradesh with Kashmir. *JBNHS* 100: 394–410.

Rahmani, A.R. (2012) *Threatened Birds of India: Their Conservation Requirements*. Indian Bird Conservation Network: Bombay Natural History Society, Royal Society for the Protection of Birds and BirdLife International. Oxford University Press. Pp. 864.

Rahmani, A.R., Islam, Z.A., Ahmad, K., Suhail, I., Chandan, P. and Zarri, A.A. (2012) *Important Bird Areas of Jammu & Kashmir*. Indian Bird Conservation Network, Bombay Natural History Society, Royal Society for the Protection of Birds and BirdLife International. Oxford University Press. Pp xii + 152.

Sahi, D.N. (1993) Wildlife Conservation sites in Kashmir Himalayas. *Tigerpaper* 20(2): 28–31.

Stattersfield, A.J., Crosby, M.J., Long, A.J. and Wege, D.C. (1998) *Endemic Bird Areas of the World: Priorities for Biodiversity Conservation*. BirdLife Conservation Series No. 7. BirdLife International, Cambridge, UK.

Sundar, K.S.G. (1999) The Sarus in Jammu, the Fulvous Whistling-duck in north Bengal and birds in Pondicherry University Campus - a reply. *Newsletter for Birdwatchers* 39(3): 41–43.

Wetlands International (2006) Waterbird Population Estimates - 4th edn. Wetlands International, Wageningen, The Netherlands. Pp. 239.

■ ■ ■

White-backed or White-rumped Vulture
Gyps bengalensis (Gmelin 1788)

ASAD R. RAHMANI

According to BirdLife International (2013), the White-backed or White-rumped Vulture qualifies as Critically Endangered because it has suffered an extremely rapid population decline, primarily as a result of feeding on carcasses of animals treated with the drug diclofenac.

Field Characters: Smallest of all Indian Gyps vultures (85 cm), it is mainly dark blackish brown with naked, thick, dark brown neck, with a white ruff at the base of the neck, dark silvery upper mandible, and conspicuous white rump (visible in flight or while spreading its wings). In flight, a broad whitish band along underside of wings is characteristic. Immatures are more brownish than black, and lack the white rump or underwing bands. Head and neck are covered with dirty white fluffy down. According to Ali & Ripley (1987), "Impossible to distinguish with certainty in the field from Long-billed Vulture, with which it is commonly associated over most of its range."

Distribution: Before the 1990s, the White-backed was probably the most abundant vulture in the world, particularly in northern India. It was also reported from Pakistan, Bangladesh, Nepal, Bhutan, Myanmar, Thailand, Laos, Cambodia, and South Vietnam, and earlier from southern China and Malaysia, but nowhere was it as abundant as in India, southern Nepal, and eastern Punjab in Pakistan. It was recorded from southeast Afghanistan and Iran where its status is currently unknown.

In Jammu & Kashmir, this species is generally restricted to the Jammu region. Intesar Suhail has recorded it from **Ramnagar** Wildlife Sanctuary (11.02.1995,

White-rumped Vulture

Place names ●
Species records ●

Srinagar
Pir Panjal Range
Dhanore Gorsian
Ramnagar WLS

Leh

Himachal Pradesh

© ISRO/NRSC; [Source : www.bhuvan.nrsc.gov.in, Data : IRS-Resourcesat-1: AWIFS]

0 65 130 260 Km

13.02.1995), and from the Bahu Fort (17.01.1997) and Bathindi (18.01.1997) areas of **Jammu** city.

Ashfaq Ahmed Zarri sighted, in November 2008, a flock of 13 birds feeding on a buffalo carcass, with around 30 Himalayan Griffon *Gyps himalayensis* and other scavengers near the Baba Ghulam Shah Badshah University, Rajouri Campus, located in Dhanore Gorsian village in **Rajouri** district. The bird was spotted in small numbers on the outskirts of Rajouri during winter every year during 2005–2011.

Pankaj Chandan and Rohit Rattan sighted seven birds during an expedition in the Pir Panjal Range in August 2010. It seems to be surviving in small numbers south of the Pir Panjal in Jammu region. However, Singh (2013) has not listed it due to its extreme rarity.

Vulture decline: The most catastrophic and rapid decline of the White-backed Vulture (and other related Gyps species) has been seen in South Asia. This decline was first reported in newspapers in the mid-1990s (see Rahmani 2008) and later confirmed scientifically at Keoladeo NP (Prakash 1999) and all over India (Prakash *et al.* 2003, Prakash *et al.* 2007). Similar steep declines were noticed in Nepal (Baral *et al.* 2005) and Pakistan (Gilbert *et al.* 2006).

For many years, virus-related disease(s) was considered the likely cause of the catastrophic decline of vultures, but in 2003, a non-steroidal anti-inflammatory drug (NSAID) diclofenac, was identified as the culprit (Oaks *et al.* 2004a, Green *et al.* 2004, Shultz *et al.* 2004). This drug is used as a pain killer for domestic livestock. If an animal dies within 2–5 days of ingestion of diclofenac and vultures feed on its carcass, they suffer renal failure causing visceral gout (Oaks *et al.* 2004a,b). The three Gyps species of vultures have declined by 97%–99% during the last 15 years. As diclofenac is widely used despite being banned, the rate of decline is nearly 50% per year of the remaining populations (Green *et al.* 2004).

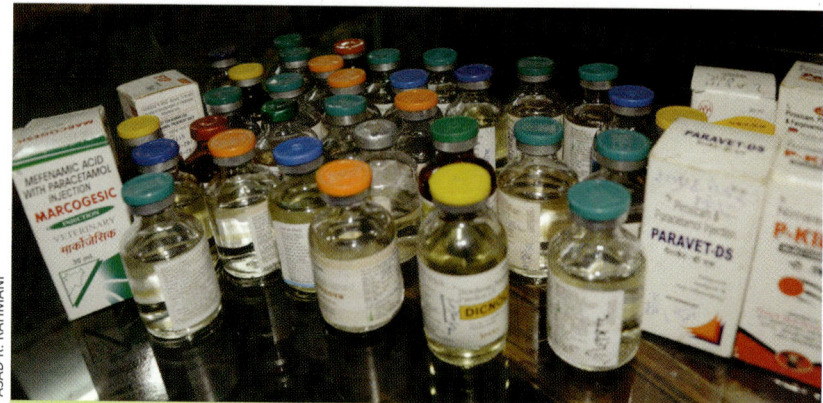

ASAD R. RAHMANI

Although officially banned since 2006 for veterinary use, diclofenac is still available in large bottles ostensibly for human use. Para vets and villagers use these multi-dose vials for livestock

Threatened Birds of Jammu & Kashmir

In the absence of vultures, most carcasses are now eaten by stray dogs whose population has shot up dramatically

ASAD R. RAHMANI

Despite 97–99% decline in the numbers of White-rumped Vulture (and other related species), it is still seen in many areas in India, although in small numbers. People have become conservation conscious and vulture sightings are reported in newspapers and bird e-groups. However, this does not mean that its population is recovering. BNHS maintains a database of all these sightings.

Ecology: The White-backed Vulture is the smallest of all Gyps vultures, and appears to be the basal from which other species of the clade of genus Gyps have diverged (Seibold & Helbig 1995, Johnson *et al.* 2006). It affects open countryside, avoiding thick forests and wooded hilly areas. Feeding as it does on large carcasses, it locates them visually, by soaring regularly on thermals, covering vast areas of hundreds of square kilometres a day. It also locates food by following other descending vultures and scavengers.

The White-backed Vulture is resident in India, with a large home range. It lives in flocks and breeds on tall trees in loose scattered colonies. However, young birds may nest solitarily. When the species was abundant, breeding colonies were found even in Delhi, Lucknow, Aligarh, Meerut, Jaipur, and many other large cities. Nests were also seen on trees along avenues, with heavy traffic below, and inside bustling towns and villages.

Thanks to the huge livestock population, and the religious taboo against the consumption of beef in India and Nepal, White-backed Vultures (and other scavengers such as Black Kite *Milvus migrans* and crows) had a regular and abundant supply of food, which would maintain their large populations. Sometimes up to 200 vultures could be seen on a cattle carcass, reducing it to a heap of bones in 20–30 minutes. Sadly, such sightings are now very rare.

Since the continuing decline of the White-backed Vulture populations from the mid-1990s, domestic animal carcasses are now mainly attended by dogs, crows, and Cattle-Egret (the last two species feed on the maggots and flies that attend a carcass). It is presumed that the population of dogs has increased correspondingly, triggering a scare of canine rabies in humans.

Threats: The White-backed Vulture, like the other two Gyps species, is in real danger of becoming extinct in another 5–10 years if diclofenac is not effectively and completely banned from the veterinary drug market. It was recorded that even if less than 1% of cattle carcasses are contaminated with a lethal dose of diclofenac, Gyps vultures will continue to die at the observed rate (Green *et al.* 2004). In a carcass sampling study conducted from May 2004 to June 2005 across 12 states (Senacha *et al.* 2008), it was found that 10.1% of the cattle carcass samples contained diclofenac sufficient to cause widespread mortality of vultures. After the ban on veterinary use of diclofenac in India in May 2006, two cattle carcass sampling studies showed that both the prevalence and concentration of diclofenac had fallen markedly 7–31 months after the implementation of the ban, with the true prevalence in the third survey estimated at 6.5% (Cuthbert *et al.* 2011). Statistical modelling of the impact of this reduction in diclofenac on the expected rate of decline of the White-backed Vulture in India indicates that the decline rate has decreased to 40% of the rate before the ban, but is still likely to be rapid (about 18% per year). Hence, constant and greater efforts to remove diclofenac from vulture food are needed to ensure future recovery of vulture populations and their successful reintroduction in the field.

Conservation measures underway: All three Gyps species are covered by Schedule I of the Indian Wildlife (Protection) Act, 1972, since 2000. They are listed in CITES Appendix II and CMS Appendix II. BirdLife International and IUCN have listed them as Critically Endangered. In 2004, the IUCN passed a BNHS/RSPB/BirdLife-sponsored resolution at its World Congress, urging all the range states to ensure effective protection of Gyps vultures. An International South Asian Vulture Recovery Plan has been developed and is being implemented in India, Nepal, and Pakistan. This Plan suggests establishing a minimum of three captive breeding centres, each capable of holding 25 pairs. Captive breeding efforts are going on, and met with success in early 2007, when two chicks hatched at a breeding centre in Pinjore, Haryana. Since then, the vultures are breeding in increasing numbers, and even double clutching and artificial incubation have been successful. In India, conservation breeding centres have been established in Pinjore in Haryana, Buxa in West Bengal, and Rani Reserve Forest near Guwahati in Assam by the RSPB and BNHS in collaboration with the state forest departments. The Central Zoo Authority of India has established four vulture breeding centres in zoos. For more details, please visit www.bnhs.org and www.rspb.org.

For **Recommendations**, see page 39.

Slender-billed Vulture
Gyps tenuirostris Gray 1844

DHRITIMAN MUKHERJEE

The most threatened vulture in the world, the Slender-billed Vulture has a very narrow distribution range, north of River Ganga up to sub-Himalaya in North India, West Bengal, and eastwards to Assam. Extralimitally, it was reported from Nepal, Bangladesh, Thailand, Malaysia, Laos, and Cambodia, and survives in Myanmar, where a small population was recorded in the Shan State (BirdLife International 2013). The Slender-billed Vulture's distribution is separate from that of the Long-billed Vulture, and its nesting habitat is also different, nests being found on tall trees, sometimes near human habitation.

Field Characters: The Slender-billed Vulture is easily distinguished from the Long-billed Vulture *Gyps indicus* by its slender jet black neck, thin elongated bill, angular black head, long legs, and toes with dark claws at all ages. The neck and head are nearly naked, with thick creases and wrinkles (prominent at close quarters). The contour feathering on its lower body is loosely textured and sparse. Due to this, a conspicuous white down patch shows on the outer side of the leg. Overall, it has an unkempt look. A prominent field characteristic is the large ear canal, unlike that of the Long-billed Vulture and other Gyps species. In size, it is closer to the Long-billed (92 cm), and much bigger than the White-backed

(85 cm). Juvenile resembles the adult, but has a black head and neck, with a hint of white down on the nape and upper neck.

Distribution: Old records reveal that the Slender-billed Vulture ranged throughout the Himalayan foothills of India, Nepal, north and central Bangladesh, Myanmar (except the north) and Southeast Asia, including Thailand, Malaysia, Cambodia, and Laos.

The species was already very rare in most of its distributional ranges in Southeast Asia in the second half of the 19th century and the first half of the 20th century, and finally it has a very limited distribution. In India and Nepal, the Slender-billed Vulture was common till the mid-1990s, but it has suffered a massive decline along with the White-backed (*G. bengalensis*) and Long-billed Vultures. This sudden crash in populations coincided with the introduction of the non-steroidal anti-inflammatory drug (NSAID) diclofenac, which is now proved to be the cause of the population decline (Green *et al.* 2004, Oaks *et al.* 2004a, Shultz *et al.* 2004, Swan *et al.* 2006a,b, Pain *et al.* 2008). Prakash *et al.* (2007) estimated its population as less than 1,000 individuals.

Its present stronghold in India is mainly in the lower Himalaya and Gangetic plain from Himachal Pradesh and Haryana in the west, to southern West Bengal, and east through Assam, and the Northeast. Rahmani (2012) has given the latest records from India. In this book, we give records from Jammu & Kashmir: Naoroji (2007), based on Ali & Ripley (1987), reported the species as a former scarce resident in Kashmir and Himachal Pradesh where it may now be an occasional summer visitor in the lower eastern regions of the state. Singh (2013) has not listed it in his recent book on the birds of Pir Panjal region. We do not have any recent specific site record from J&K. Therefore, a thorough survey is required of the state.

Ecology: Like other Gyps species in India, the Slender-billed Vulture affects open dry country, often near human habitations, mainly at carcass dumps where it feeds, along with its congenerics, on carrion. Despite its hooked bill and sharp claws, this vulture does not kill its prey, preferring to feed on carcasses of large or medium-sized ungulates. It tolerates human presence, coming to breed near villages on tall trees. In Southeast Asia it was found generally in the lowlands, in partly wooded country.

It is gregarious — feeding, roosting, and resting in large loose flocks, often with other vulture species. It breeds in winter, from November to May. The nest is built on large tall trees, from 8–12 m high, often in loose colonies. A single egg is laid, and both parents help in incubation and raising the chick.

Threats: The main threat to the Slender-billed and other species of Gyps vultures in Asia comes from the veterinary use of the non-steroidal anti-inflammatory drug (NSAID) diclofenac. A detailed account of this threat is given in the section on White-backed Vulture *Gyps bengalensis*. Other threats are changes in disposal method of dead livestock, unintentional poisoning, and accidental collision with vehicles or trains, but these are probably minor.

Conservation measures underway: Like the other two Gyps vultures of India, the Slender-billed Vulture is listed in Schedule I of the Indian Wildlife (Protection) Act, 1972, since 2000. It is also listed in CITES Appendix II and CMS Appendix II. BirdLife International (2013) and IUCN have given its status as Critically Endangered. In 2004, the IUCN passed a BNHS/RSPB/BirdLife-sponsored resolution in its World Congress urging all the range states to work in cooperation for the effective protection of Gyps vultures. A South Asian Vulture Recovery Plan (Anon. 2004) was developed and is being implemented in India, Nepal, and Pakistan (this species is not found in Pakistan, but the recovery plan applies to all three Gyps species).

RECOMMENDATIONS FOR TWO GYPS SPECIES OF VULTURES FOUND IN JAMMU & KASHMIR

As the Government of India has officially banned veterinary use of the drug diclofenac, it is not available in the market. However, formulations of diclofenac for human use are widely used illegally on livestock despite the ban.

(1) The central government with the help of J&K state government should strictly implement the ban on veterinary use of diclofenac, including the use of human formulations of diclofenac on livestock.

(2) We recommend that all competent organisations and agencies seriously implement programmes to raise awareness about the hazards of diclofenac poisoning of vultures among the general public and especially among major stakeholders, including farmers, graziers, veterinarians, pharmacists, staff of government and state wildlife and agricultural agencies, and religious and other groups which place special value on the continued existence of vultures.

(3) Appropriate authorities must undertake thorough evaluation of pharmaceuticals likely to be used in place of diclofenac to ensure that they are not toxic to vultures and other scavengers. The Ministry of Environment and Forests, Government of India should insist that any new NSAID for veterinary use should be safety-tested on vultures before it is introduced in the market.

(4) Other factors not responsible for the recent catastrophic declines may assume increasing significance in future, as the depleted populations fall still further. So, we recommend that all such factors including poisoning of cattle carcasses, injury or death of vultures due to kite flying, and disturbance to nests should be stopped by appropriate conservation measures.

(5) The Ministry of Environment and Forests should ensure that funding does not become a constraint in running the Vulture Conservation Breeding Programme, which is of paramount importance to ensure the survival of these valuable scavenging birds that play a vital role in quick disposal of carcasses, and consequently in human environmental management.

(6) Thorough scientific survey of all vultures should be conducted in the state, followed by annual population monitoring.

SAVING ASIA'S VULTURES
FROM EXTINCTION

A consortium Saving Asia's Vultures from Extinction (SAVE) was launched in February 2011 in Delhi and Kathmandu to provide a strategic framework through which the unprecedented problem and threat to South Asian Gyps vultures could be addressed across national boundaries. It provides a clear scientifically based outline of the priorities that need addressing to conserve the most threatened species, and also a recognised channel for supporters to ensure that resources are used to address those priorities.

SAVE consists of six core members: Bird Conservation Nepal, Bombay Natural History Society, International Centre for Birds of Prey (UK), National Trust for Nature Conservation (Nepal), the Royal Society for the Protection of Birds (UK), and WWF Pakistan, and a growing number of project and research partners including the Indian Veterinary Research Institute. Professor Ian Newton, world renowned raptor expert, agreed to take the chair for the first four years; and there are two main subcommittees that help drive the research, field actions, and advocacy that is needed.

Vulture conservation efforts in India are showing the first signs of success thanks to the initiative of the Indian government (led by the Ministry of Environment and Forests) in banning veterinary formulations of diclofenac in 2006, which has had an important impact in slowing the declines. The breeding programme includes three BNHS-run centres in Haryana, West Bengal, and Assam with the support of the respective state government Forest Departments. The Central Zoo Authority is supporting further breeding facilities to extend these efforts at five more zoos. A Regional Steering Committee has been established through an IUCN initiative in 2012 with National Vulture Recovery Committees being set up in Pakistan, India, Nepal, and Bangladesh, which will be an important forum for delivering the further measures required to conserve vultures in the subcontinent.

A website, www.save-vultures.org, provides full details and more information, as well as all key Asian vulture publications available for download, the manifesto, and most importantly a donations button where supporters can help ensure that resources are available to support these vital efforts.

www.save-vultures.org   **Chris Bowden, SAVE Programme Manager**

Egyptian Vulture
Neophron percnopterus (Linnaeus 1758)

VISHWATEJ PAWAR

The Egyptian Vulture is perhaps the most widespread vulture of the Old World, with isolated resident populations in the Cape Verde and Canary Islands off the West African coast, north Africa, Ethiopia, and east Africa, isolated populations in Angola and Namibia, southern Europe, the Mediterranean, the Middle East, Central Asia to India and Nepal. In its wide range, it is declining rapidly, therefore BirdLife International (2013) lists it as Endangered. Also called Pharoah's Chicken, it is a long-lived and slow breeding bird with very few predators during its adult phase, therefore any decrease in breeding or increase in adult mortality, as seen in southern Europe (>50% over the last three generations, i.e. 42 years) could spell doom for this species. India, where good populations used to be present 20 years ago, has also registered a sharp decline. BirdLife International (2013) estimates that the total world population is between 13,000 and 41,000 mature individuals.

Field Characters: The Egyptian Vulture has a unique, shabby, scruffy appearance. It is a small kite-like bird with naked head and short, entirely feathered neck. Adult dirty white with black flight feathers. Juvenile dark, with pale vent and tail. The face is bare, yellow in adult and brown in juvenile.

Distribution: The Egyptian Vulture has a very wide range extending from Africa, southern Europe, the whole of the Middle East, Iran, Afghanistan, and Pakistan, to India and Nepal. For details of population figures, see BirdLife International (2013).

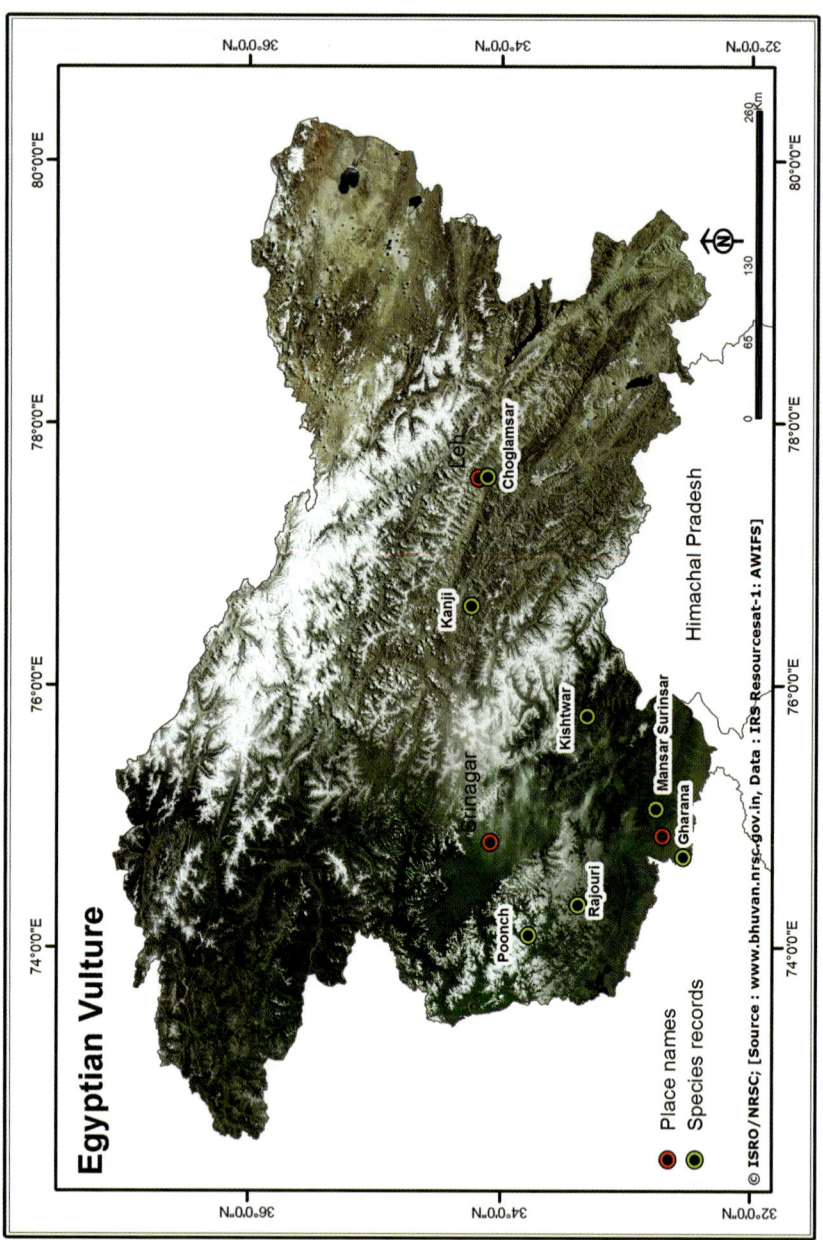

Egyptian Vulture

Place names
Species records

© ISRO/NRSC; [Source : www.bhuvan.nrsc.gov.in, Data : IRS-Resourcesat-1: AWIFS]

Himachal Pradesh

Leh
Choglamsar
Kanji
Kishtwar
Srinagar
Poonch
Rajouri
Mansar Surinsar
Gharana

It is found all over India, from the plains to *c.* 2,500 msl, sometimes very close to human habitation, but its numbers are decreasing. Still widespread in India, it is frequently seen in Uttarakhand, Rajasthan, Gujarat, Uttar Pradesh, Madhya Pradesh, Chhattisgarh, Maharashtra, and decreasingly so in south India. Rahmani (2012) has given important sight records of this species. Here we give relevant records from Jammu region. The bird has been seen frequently in small numbers near **Gharana** Wetland Reserve by Ashfaq Ahmed Zarri, Pankaj Chandan, Pushpinder Singh Jamwal, and Rohit Rattan. It was also recorded on three different visits to the **Mansar Surinsar** Wildlife Sanctuary in the winter of 2009, 2010, and 2011 (Ashfaq Ahmed Zarri). The species has been sighted recently in various other locations in **Rajouri** and **Poonch** districts (Ashfaq Ahmed Zarri & Pankaj Chandan). A few birds were seen near **Chandimarh** village in Poonch district during an expedition in the Pir Panjal Range by Pankaj Chandan. The species can be easily seen along River Tawi and the outskirts of **Jammu** city. Thus, although the overall population seems to be declining in J&K, a small population is still surviving in the Jammu region, south of Pir Panjal range.

Pfister (2004) reported sighting a single bird from **Choglamsar** at 3,300 msl in Leh, Ladakh. More recently, Nisa Khatoon of WWF-India (*pers. comm.* 2012) sighted two birds in June 2012 in **Kanji** Wildlife Sanctuary in Kargil district, Ladakh at 3,857 msl.

According to Brigadier Rajendra Singh (*pers. comm.* 2013) it has been recorded and photographed from the following places: 5 to 7 km downstream of **Poonch** town on the banks of the Poonch river (February 27, 2012); east of **Jammushid** village, **Poonch** district (April 13, 2012); photographed on a ridge east of **Jammushid** village (September 26, 2012); and photographed near **Kishtwar** town (April 15, 2012). On all these occasions, two or three vultures were seen. They were also seen on numerous occasions near **Jammushid** village, sometimes from such a distance that they could not be photographed; they appeared to be breeding there.

Ecology: The Egyptian Vulture can be seen sauntering around villages and nomad camps, looking for carrion, offal, garbage, and human excrement. It opportunistically picks up crickets, frogs, and alates of emerging termites. It has a narrow long beak, which helps it in tearing off small pieces of meat through narrow spaces between bones, where larger-beaked vultures cannot reach.

The Egyptian Vulture is usually solitary or found in pairs with juveniles, but on good feeding sites (e.g., Jorber carcass dump, Bikaner) 1,000–2,000 are seen in winter (Vibhu Prakash *pers. comm.* 2010). It roosts singly or in small groups, generally on tall trees, but electric pylons are frequently used where tall trees are absent. Although it is mostly resident and seen around its usual haunts throughout the year, the northern populations undertake short to long distance migration, as conditions become unsuitable during winter. It feeds on dead animals, but can also kill stranded fish and turtles, and other small prey. It was seen killing Checkered Keelback *Xenochrophis piscator* in Keoladeo NP, Bharatpur (Prakash & Nanjappa 1988).

It mainly nests on cliffs, rocky outcrops, ledges of occupied buildings, abandoned forts and ruins, but occasionally on tall trees, where its preferred nesting habitat is not available. A single egg is laid and both parents share all parental duties, including incubation.

Threats: In its vast distributional range, threats vary from country to country and region to region. In some areas, loss of wild ungulate populations and hence reduced food supply is the main threat, while in some countries antibiotic residues in cattle carcasses could be the major threat. Death by hitting powerlines is an additional risk in some European countries where the population is anyway small. In India, the annual rate of population decline was 35% during 2000–2003, and the population in 2003 was estimated to be 20% of that in the early 1990s (Cuthbert *et al*. 2006).

In India, the main threat could be diclofenac poisoning by feeding on contaminated cattle carcasses, as in the Gyps vultures. It is probable that other NSAIDs used in livestock could be fatal for these vultures also. Earlier, when Gyps vultures were in abundance, they would not allow the Egyptian Vulture to feed on the internal organs such as lungs and liver of a carcass, but now with the near total disappearance of Gyps vultures from the Indian subcontinent, the Egyptian Vulture has a greater chance to feed on such internal organs that contain more diclofenac than the muscles and tendons on which this bird fed in the past. Thus the risk increases by loss of competition.

Conservation measures underway: The Egyptian Vulture is listed in Schedule IV of the Indian Wildlife (Protection) Act, 1972. In India, it still occurs in numerous PAs and IBAs. The veterinary use of diclofenac has been totally banned by the Indian Government since 2006. Regular surveys by BNHS, funded by RSPB, are ongoing in India.

RECOMMENDATIONS

Rahmani (2012) has given India-specific recommendations to which we have added state-level recommendations:

(1) Total ban on veterinary use of diclofenac should be implemented in the whole country. Human use diclofenac should not be sold and administered for veterinary use.
(2) Studies on the impact of other NSAIDs on Egyptian Vulture.
(3) All India surveys on a regular basis, to study the population trend. Conduct surveys in Jammu region to know its distribution, status, and specific threats.
(4) Initiation of an environmental education campaign in rural areas about the importance of vultures.
(5) Ecological and behavioural studies on this species.
(6) Study of its movement through satellite tracking, to map its home range in breeding and non-breeding seasons, and also the dispersal of juveniles.

White-headed Duck
Oxyura leucocephala (Scopoli 1769)

BirdLife International (2013) has included this species in the Endangered list, based on mid-winter counts (organised by Wetlands International) which indicate that the population of this species has undergone a very rapid decline.

Field Characters: The White-headed Duck, as the name indicates, has a white head but only in the adult male, with a small black cap and blue bill, swollen at the base. Body of male and female are chestnut brown, with a pointed tail which is often held erect. Females have dark face-stripes and a greyish bill. The species prefers large freshwater-bodies, lakes, and brackish lagoons. It was also known as White-headed Stifftail due to its habit of keeping its tail upright when swimming (Ali & Ripley 1987).

Distribution: The White-headed Duck *Oxyura leucocephala* occupies a large Palaeartic range that stretches from Spain and Algeria in the west, to Mongolia, western China, and India in the east. This distributional range is highly fragmented, and it apparently became extinct as a breeder in Morocco, central Europe, and Palestine in the 20th century (Green & Anstey 1992, BirdLife International 2001). It breeds in western Mongolia, and parts of eastern Russia, and there are reports of its breeding in western China too. It is a non-breeding visitor to Pakistan, and (in very small numbers) to northern India (Ali & Ripley 1987).

For details of distribution and population, see BirdLife International (2013) Species Factsheet. Sightings of White-headed Duck are regularly reported outside the breeding and winter ranges, but these extralimital records could be of escaped individuals from wildfowl collections and zoos.

White-headed Duck

Place names
Species records

© ISRO/NRSC; [Source : www.bhuvan.nrsc.gov.in, Data : IRS Resourcesat-1: AWIFS]

Himachal Pradesh

Wular Lake
Kashmir Valley
Srinagar
Jammu
Leh

Nearly 130 years ago, White-headed Duck was reported from Wular Lake. It could still be visiting some of the wetlands of the valley: regular waterfowl monitoring is required

In India, it is reported from Jammu & Kashmir, Punjab, Haryana, Uttar Pradesh, Delhi, Gujarat, and West Bengal. For details of sight records, see Rahmani & Islam (2008). Rahmani (2012) has mentioned three recent records from Harike in Punjab, Amakhera in Uttar Pradesh, and Gajaldoba Barrage in West Bengal.

In **Jammu & Kashmir**, it was recorded from **Srinagar**. Oliver (1919) shot one bird on November 7, 1918, and three more on November 23, 1918. Unwin (1897) recorded it from the Kashmir Valley as late as mid-April (as a very uncommon bird). According to Walter Lawrence (1895), the duck is 'very rare' in Kashmir and 'six specimens were shot on or near the **Wular lake** by a sportsman in the hard winter of 1890–1891.' We do not have any record in recent years. Ever since the establishment of a Wetland Division under the Department of Wildlife (Protection) in J&K, efforts are being made by field staff during surveys to locate this species, but not even a single bird has been spotted in any wetland of Kashmir in recent years (Imtiaz Lone, Wildlife Warden, *pers. comm*. 2012).

Ecology: The ecology and breeding biology of the White-headed Duck have been extensively studied (Green & Hughes 1996, Green *et al*. 1999, Johnsgard & Carbonell 1996, Sánchez *et al*. 2000). The following description is based mainly on the above references:

Although it is highly gregarious in the wintering areas, with records of more than 10,000 individuals in winter sites, during the breeding season pairs select well-vegetated shallow eutrophic waterbodies where nests are built in dense vegetation. The female lays 6–13 eggs. It is a diving duck, and in many ways reminiscent of the grebes. It lies very low on the water, like a cormorant, showing only its head, a small portion of rump, and the stiff, pointed tail usually cocked vertically (Wright & Dewar 1925). Its food consists mainly of midge larvae (Chironomidae) and other aquatic invertebrates such as amphipods, isopods, and polychaetes (especially in coastal wintering sites). Seeds and the vegetative parts of *Potamogeton* spp., *Ruppia* spp., and other aquatic plants are also consumed. A bird killed in India was found to have only vegetable matter in its stomach (W.A. Whitehead, *JBNHS* 35: 212).

Threats: Like other waterfowl, the White-headed Duck is affected by drainage and pollution of its wetland habitat, increased hunting pressure, and disturbance during breeding. Unsustainable use of water resources and drought conditions are other problems. These threats are likely to be exacerbated by the effects of global climate change. In addition, the European population, particularly in Spain, faces competition and hybridisation with the more aggressive Ruddy Duck *Oxyura jamaicensis*, an invasive species from America. Attempts are being made to locally eradicate this invader to maintain the authentic gene pool of the White-headed Duck (BirdLife International 2013).

Conservation measures underway: The species is legally protected in many range countries, including India. It is included in Schedule IV of the Indian Wildlife (Protection) Act, 1972. It is listed in CITES Appendix II and CMS Appendix I. A conservation programme in Spain has resulted in a significant population increase (BirdLife International 2013).

RECOMMENDATIONS

As it is a scarce migrant in India, not much can be done here except prevention of poaching and poisoning of all waterfowl. The IBAs and PAs where this bird has been reported should be better protected, particularly their water quality.

Western Tragopan
Tragopan melanocephalus (Gray 1829)

K. RAMESH

The Western Tragopan, an endemic bird of the western Himalaya in India and Pakistan, is considered Vulnerable by BirdLife International (2001, 2013) because its sparsely distributed, small population is declining and becoming increasingly fragmented in the face of continuing forest loss and degradation throughout its restricted range. Its numbers are now estimated to be not more than 2,500 to 3,500 birds in the wild (Sanjeeva Pandey *pers. comm.* 2010).

Field Characters: The Western Tragopan (68–73 cm) is a brilliantly coloured, beautiful bird, particularly the male, which is dark with white ocelli, crimson hindneck patch, orange foreneck, reddish facial skin, blue throat patch, a small dark occipital crest, and white uppertail-coverts with black tips. Underparts are black, ocellated with white irregularly smeared with red, flanks and abdomen mottled with brown and black. It can be confused with Satyr Tragopan *Tragopan satyra*, but is much darker, with different facial and body coloration. However, the females of both the species are indistinguishable in the field. Female is smaller, greyer, and finely marked, and without the red and blue facial markings of the male. For details of male, female, first-year male, chick, and eggs, see Delacour (1951).

Western Tragopan

Limber WLS
Srinagar
Tatakuti
Dehra Gali
Kishtwar
Leh

Himachal Pradesh

Place names
Species records

© ISRO/NRSC; [Source : www.bhuvan.nrsc.gov.in, Data : IRS-Resourcesat-1: AWIFS]

Distribution: The Western Tragopan is one of the rarest pheasants of the world, with a narrow distribution zone in the temperate forests of the Northwest Himalaya. It has been reported from Swat catchments in the North West Frontier Province (NWFP), Pakistan, through Kashmir, Himachal Pradesh, and up to west Uttarakhand, mostly between 2,400 and 3,600 msl, but even as low as 2,000 msl in winter. It prefers dense undergrowth and montane bamboo thickets in undisturbed temperate and subalpine oak, coniferous, and mixed forests (Ali & Ripley 1987, Islam & Crawford 1987, Fuller & Garson 2000, BirdLife International 2001).

In India, it is found in three states: Uttarakhand, Himachal Pradesh (which has the largest population), and Jammu & Kashmir (very few recent records). Rahmani (2012) has described its latest distribution records. Here we give records valid for Jammu & Kashmir.

In J&K, the bird has historically been distributed only along the Pir Panjal Range (which borders the valley with Himachal) but not towards the inner Greater Himalayan Range (Intesar Suhail & Khursheed Ahmed *unpubl.* 2011). It has recently been recorded from Baranari-Kalamund and Hingli localities of **Kalamund-Tatakuti** Sanctuary, and Gide Nuk and Pathra localities of **Khara Gali** Area in **Poonch** (Riyaz Ahmed *pers. comm.* 2011). In May 2012, Riyaz Ahmad heard 12 males in Kalamund-Tatakuti area and eight from Khara Gali area of Poonch Pir Panjal. Hillaluddin (*pers. comm.* 2013) sighted four birds during survey in 2006–2007 in Kishtwar National Park (Jammu).

It was never observed in a seven-year study at Overa Sanctuary in Kashmir (Price *et al.* 2003) nor was it recorded during surveys in Dachigam, Overa, and Naganari in April–May, 1981 (BirdLife International 2001). It clearly has a disjunct distribution in the Himalaya, only occurring in some patches. Islam & Rahmani (2004) in *Important Bird Areas in India* have mentioned that it was reported as secondary information from Dehra Gali Forest (**Poonch**, **Rajouri**), **Kishtwar** NP (**Doda**), **Lachipora** WLS (**Baramulla**), and **Limber** WLS (**Baramulla**). Many such site records need confirmation. This species has also been reported earlier from **Wular** catchment (Knox & Walters 1994); **Lolab Valley** (Baker 1921); **Limber** Wildlife Sanctuary (Kaul 1989, Qadri *et al.* 1990, Javed 1992, Akhtar *et al.* 1994); and **Kishtwar** National Park (Ward 1906–1908). More recently, it was recorded during the surveys carried out by the Department of Wildlife Protection during 1992–93 (two individuals), in 1998–99 (four individuals) and in 1999–2000 (five individuals) in **Kishtwar** National Park in Jammu region.

As this is a very conspicuous bird and much prized by hunters, it has many names such as Sonalu for male, Solalee for female (Kashmir), Fulgar for male, Fulgari for female (Chamba, Himachal), Pyara (Kinnaur, Himachal), Sing Monal (Pahari), and Jewar (Garhwal). It is called Jujurana or the King of Birds in the Kulu-Manali area of Himachal. In the erstwhile Bushahr state (Rampur Bushahr), it was known as Jyazi.

The Western Tragopan has a narrow distribution zone in the temperate forest of north-west Himalaya, mostly between 2,400 and 3,600 msl

Ecology: It lives in pairs or family groups, is generally timid, and forages in thick vegetation. However, wherever unmolested and in remote areas with little human presence, it can be seen foraging on slopes and forest openings in the morning and evening. It feeds on the ground, on fresh leaves, fallen fruits and seeds, shoots of ringal bamboo, berries, and insects. The nest is made either on the ground in thick vegetation, or on boughs or in cavities of large trees, not very high. Clutch size is 3–6 (roughly three, according to Ramesh *et al.* 2008), incubated by the drab-coloured hen for about 26 days, while the cock stays around the nest to provide protection. Both sexes help in raising the chicks. Breeding season is May–June, when the male shows a bizarre and curious display, strutting around, and flashing small, fleshy, blue, erect 'horns' on either side of the crown, hence its name Horned Pheasant. The display has been described in detail by Delacour (1951), Ali & Ripley (1987), and Roberts (1991).

Like all pheasants, the Western Tragopan is highly territorial and makes a nasal, wailing *khuwaah* call, repeated 7–15 times during the breeding season. When disturbed or agitated, it calls *waa waa waa* (Sanjeeva Pandey *pers. comm*. 2010). Ramesh *et al.* (2008) stated that the Western Tragopan appears to show site fidelity, at least in the breeding season.

A great deal of conservation research has been done in the last 20 years on this much-loved species, particularly in the Palas Valley of Pakistan (e.g., Islam & Crawford 1987, Bean *et al.* 1994) and India (Pandey 1995, Ramesh 2003, Jandrotia *et al.* 1995, Jandrotia *et al.* 2000). Recently, Ramesh *et al.* (2008) studied a radio-tracked female Western Tragopan for six months and gathered valuable information on its ecology (despite low sample size). They found that the bird showed preference for high tree cover, thick undergrowth of montane bamboo, high litter cover, and perennial waters. The estimated home range was 31.6 ha in winter, 20.5 ha in summer and 4.7 ha in autumn. Ramesh *et al.* (2008) state that the radio-tagged bird proved important to substantiate the general claim of dense undergrowth, such as high altitude montane bamboo, being an important cover species for Western Tragopan in the Great Himalayan NP. In other parts of its range, the important cover species is *Viburnum* sp. Ramesh *et al.* (2008) also point out that the intensity of the use of bamboo patches was largely overlooked by conventional studies using trail monitoring.

Threats: Habitat degradation and fragmentation, and hunting are the main direct threats throughout its range, in various degrees from one area to another. Indirect threats include heavy grazing pressure, and collection of fuel wood, wild mushrooms, and medicinal plants, which create huge disturbance, particularly in the summer when the bird is nesting (Gaston *et al.* 1983).

This tragopan has a huge international demand but no collection has taken place in recent years in India. Earlier data shows that 200 or more Western Tragopan were exported but none survived. Compared to other pheasants, it is a difficult species to acclimatise and maintain in captivity.

Conservation measures underway: Western Tragopan is listed in Schedule I of the Indian Wildlife (Protection) Act, 1972. It is also listed in CITES Appendix I, and international trade of the live bird or its products (feathers) is illegal. It occurs in many national parks in Pakistan and India, and a well-known initiative involving local people in the Palas Valley, Pakistan has shown positive outcome under the Himalayan Jungle Project. In India, it is found in many sanctuaries and IBAs (Islam & Rahmani 2004). A Western Tragopan conservation breeding programme is being undertaken at Sarahan Bushahr Pheasantry by the Wildlife Wing of Himachal Pradesh Forest Department, in collaboration with the World Pheasant Assocation International, which has potential for future release of parent-reared offspring to augment/restock local wild populations.

RECOMMENDATIONS

(1) Accord Western Tragopan flagship species status for the conservation of moist temperate forests and other pheasant species.
(2) Develop monitoring methods, and monitor key populations regularly.
(3) Conduct further surveys, particularly in Jammu & Kashmir along the Pir Panjal Range, and map known sites.
(4) Conduct ecological studies on radio-tagged birds.

Cheer Pheasant
Catreus wallichii (Hardwicke 1827)

KALYAN SINGH SAJWAN

BirdLife International (2001, 2013) justifies the inclusion of the Cheer Pheasant in the Vulnerable category as it has small naturally fragmented populations, because it lives in small patches of successional grassland. Human population pressure, grazing pressure from livestock, hunting, and changing patterns of land use are resulting in its decline within this habitat.

Field Characters: Unlike other pheasants, there is not much plumage difference between the male (90–118 cm) and female (61–76 cm), except that the female is much smaller, duller, and more heavily marked, with a smaller crest. The Cheer Pheasant also lacks the colour and brilliance of most pheasant species. Overall, it is a grey, brown, and buff bar-tailed pheasant with long backwardly-pointed blackish-brown crest, and crimson orbital patch or facial skin. The male has largely plain pale greyish upper neck and throat, and clear, dark barring on upperparts. Lower part chiefly buffy-white, conspicuously barred on lower breast and flanks. Centre of abdomen blackish. Tail long and pointed in both sexes. Immature male like the female, less barred and lacks the crest.

Voice is a loud *chir-a-pir chir-a-pir chir chir-chirwa chirwa* and high, piercing *chewewoo* notes, interspersed with short *chut* and harsh staccato notes. Hence

Cheer Pheasant

Limber WLS
Tatakuti
Srinagar
Vaishno Devi
Jammu
Leh

Himachal Pradesh

Place names
Species records

© ISRO/NRSC; [Source : www.bhuvan.nrsc.gov.in, Data : IRS-Resourcesat-1: AWIFS]

its name Chir or Cheer Pheasant. In addition, various clucks and chuckles, to express contentment, alarm and other emotions, are also emitted (Ali & Ripley 1987). Rasmussen & Anderton (2005) say "Calls include a long, rhythmic series of staccato, brassy *scritch, scritch* clacks or rapid *skew-it-churt*."

Distribution: The Cheer Pheasant is found mainly between 1,400 to 3,500 msl in the Western and Central Himalaya, from north Pakistan, through Kashmir into Himachal Pradesh and Uttarakhand, and east to central Nepal. However, it has been seen as low as 950 m at Birahi, Chamoli district (Bisht *et al*. 2007) and as high as 4,545 m in Uttarkashi district (Ghosh 1997).

BirdLife International (2001) has given its detailed historical and present-day range from hunting and sighting records, and museum specimens. Here we give recent records from Jammu & Kashmir.

In Kashmir, the Cheer Pheasant is found only in the **Kazinag** hills (Limber, and perhaps Lachipora), the strength of these populations is not known. In the Jammu region, it occurs in the **Trikuta** hills in Jammu and may also be found in areas of **Kishtwar** and **Bhadarwah**, although this has not been confirmed so far. It may also occur in the Rajouri hills. Riyaz Ahmad (*pers. comm*. 2013) sighted four birds and heard 10 birds in Kalamund-Tatakuti, and heard four calls in Khara Gali of Poonch Pir Panjal. Hillaluddin Ahmad (*pers. comm*. 2013) has sighted and photographed one bird near Mata Vaishno Devi temple, Reasi district in 2012.

Ecology: It is a resident bird of the Western Himalaya where it inhabits precipitous, craggy, and rocky hillsides with scrub and stunted trees, dissected by wooded ravines or with some scrub and grass cover. It frequents altitudes of 1,200 to 3,350 msl. There is a record of its occurrence at 4,545 msl, near Dharoadhari in Uttarakhand (Ghosh 1997). It strongly favours early successional stages of forest regrowth, often created in the first place by traditional grass cutting and burning regimes (see Garson *et al*. 1992). As these sites are generally near human habitations, it suffers from heavy hunting pressure. Moreover, in some protected areas where grass cutting and tree lopping have been curtailed, the habitat soon becomes unsuitable for this species (Sathyakumar *pers. comm*. 2010).

Bisht *et al*. (2007) found that the Cheer Pheasant prefers the slopes that do not receive direct sunlight most of the day, as they have comparatively soft soil which enables it to dig out tubers and roots for food.

Breeding season is from April to June, extending to September, with clutch sizes of 6–12 eggs. Nest is a scrape on the ground in the dense undergrowth or behind a boulder on steep hillside in chir or oak forest, concealed by overhanging grasses and/or branches. Although incubation is done by the hen alone for about 26 days, the cock stands guard around the nesting area. The species is possibly monogamous and the cock is reported to assist the hen in the care of the young (Delacour 1951). Its food consists of roots, berries, fruits, insects, and other small invertebrates. Where not molested, it can be seen foraging in crop fields early in the morning and late evening.

Population estimates: BirdLife International (2013) estimates total world population in the wild to be between 4,000 and 6,000 in a total area of 27,500 sq. km. This population is decreasing. M.S. Bisht (*pers. comm.* 2010) believes that it could be more, as it is very difficult to estimate because of (i) its shy habits and occurrence in many inaccessible areas, (ii) no intensive study being made so far along whole distributional limits, and (iii) all previous reports being based either on single sighting or on survey of small areas for a few days only. However, recent surveys by R. Kalsi (*pers. comm.* 2010) of the previously surveyed sites in Himachal Pradesh showed that Cheer Pheasant populations have declined considerably, and they have disappeared altogether from some sites. There has been considerable decline in Chail and Majathal WLSs (both in Himachal), which were considered strongholds of the species. It is feared that the same condition prevails in Uttarakhand as well. Therefore, at present there could be 3000–4000 Cheer in the wild (Rajeev Kalsi *pers. comm.* 2010).

Threats: Hunting and habitat modification are the major threats to this species.

Conservation measures underway: This species is listed in Schedule I of the Wildlife (Protection) Act, 1972, and also Appendix I of CITES. It is legally protected in all the range countries (Pakistan, Nepal, and India). It occurs in a number of protected areas in India, Nepal, and Pakistan. It is also found in many PAs/IBAs in India and is a well-studied species, with many status surveys done in all the three countries where it occurs.

RECOMMENDATIONS

(1) Conduct surveys in the state to find out its distribution and status in different areas.

(2) All previously reported Cheer sites must be revisited and surveyed for Cheer presence, absence, and population status, using new techniques of population estimation and analyses.

(3) In the specific areas where we have past and present data on Cheer relative abundance (qualitative or quantitative), habitat management could be carried out on an experimental basis. For instance, areas that have not been grazed or have been unburnt for several years could be treated with controlled fire or moderate grazing and monitored.

(4) Strict control on hunting and poaching through enforcement and public awareness should be ensured.

(5) Provide protection to non-protected IBAs through involvement of local communities, and identify sites for community reserves.

Lesser White-fronted Goose
Anser erythropus (Linnaeus 1758)

According to BirdLife International (2001, 2013), the Lesser White-fronted Goose is listed as Vulnerable in the IUCN Red List because its key breeding population in Russia has suffered a rapid population reduction and equivalent decline is predicted to continue. There is no recent record from Jammu & Kashmir.

Field Characters: The Lesser White-fronted Goose *Anser erythropus* is closely related to the Greater White-fronted Goose *A. albifrons*. It is small (53 cm), even smaller than some ducks such as the Indian Spot-billed Duck *Anas poecilorhyncha* (61 cm). It has a diagnostic white patch on the forehead, sloping forehead and small head, short pink bill, and yellow eye-ring. Overall it is grey-brown, with blotchy black bands on lower breast and belly. Due to its size and characteristic features, it is not difficult to identify. It walks faster than *A. albifrons* and is consequently often found at the front of feeding flocks (BirdLife International 2013). An attractive species, it is often kept in wildfowl collections, from where escapes do occur: individuals seen in summer, or in the company of other feral geese, are likely to be of captive origin.

Distribution: The Lesser White-fronted Goose breeds in the taiga and tundra zones of northern Eurasia, from Arctic Europe to northeastern Siberia (Russia), and winters primarily in southeastern Europe, around the Black and Caspian seas, at the lower Euphrates in Iraq, and in the lowlands of eastern and southern China (Madsen 1996). It has been recorded outside the breeding season in Japan, South Korea, mainland China, Taiwan, Pakistan, India, and Myanmar, with unconfirmed reports from Bangladesh. Most of the Asian population is believed to winter in the lower Yangtze river floodplains of China, where Barter *et al.* (2005) counted almost 17,000.

Lesser White-fronted Goose

Wular Lake

Srinagar

Leh

Himachal Pradesh

● Place names
◉ Species records

© ISRO/NRSC; [Source : www.bhuvan.nrsc.gov.in, Data : IRS-Resourcesat-1: AWIFS]

ASAD R. RAHMANI

The Wular Lake, where this species was reported more than 100 years ago, has now changed due to expansion of agriculture and tree plantation

In India, it is a rare and sparse winter visitor. It was recorded sporadically from **Kashmir**, Uttar Pradesh, Bengal, Bihar, and Assam (Ali & Ripley 1987). For historical records, see BirdLife International (2001) and Rahmani & Islam (2008). Recent records in India are given by Rahmani (2012). In **Jammu & Kashmir**, it has been recorded from the **Wular Lake**, where it was apparently shot in 1907 (Ward 1906–1908). Baker (1908) has recorded it from four unspecified localities.

Ecology: As it is a marginal species in the state, we are not describing its ecology in detail. The Lesser White-fronted Goose breeds mainly within the forest zone at the southern edge of the Arctic Circle and in the tall-shrub tundra belt, and after breeding it migrates north to the Arctic coastal lakes in summer where it is seen in non-breeding plumage. It winters in open treeless tracts, including semi-arid salt steppes, meadows, pastures, and sometimes cropfields, and roosts in reed beds and rushes, or on water or the banks of lakes and rivers; it rarely visits marine waters (Cramp & Simmons 1977).

Threats: Illegal hunting and disturbance to its breeding areas are the biggest threats. In India, it faces the common threats of hunting, poisoning, drainage of wetlands, and pollution.

Conservation measures underway: It is listed in Schedule IV of the Indian Wildlife (Protection) Act, 1972, so hunting and shooting is illegal. It is listed in CMS Appendix I and II.

RECOMMENDATIONS

As it is a marginal species to the state, we do not have any specific recommendation. We have to prevent poaching and hunting of all such species and protect key areas.

Marbled Duck
Marmaronetta angustirostris (Mènètriés 1832)

There are a few old records of the Marbled Duck from the Kashmir Valley but no recent record. Therefore it is considered to be a marginal species in the state. BirdLife International (2013) justifies keeping this species in the Vulnerable category as it has suffered a rapid population decline, evident in its core wintering range, as a result of widespread and extensive habitat destruction.

Field Characters: A small duck of about 48 cm, with an overall coloration of grey-brown flecked with creamy-brown. It has a dark eye-patch and broad eye-stripe from eye to nape. A slight nuchal crest is present. No speculum or wing mirror, unlike teals. Ventral parts are sullied white, more or less barred transversely with brown (Ali & Ripley 1987). Female slightly smaller. Juveniles are similar to adults but with more off-white blotches. In flight, the wings look pale without a marked pattern, and lack speculum on the secondaries. Pale blotches on the mantle and wing coverts give a marbled effect to the plumage, hence the name.

Distribution: The Marbled Duck has a fragmented distribution in the western Mediterranean (Spain, Morocco, Algeria, Tunisia, wintering in north and sub-Saharan west Africa), the eastern Mediterranean (Turkey, Israel, Jordan, Syria, wintering south to Egypt) and western and southern Asia (Azerbaijan, Armenia, Russia, Turkmenistan, Uzbekistan, Tajikistan, Kazakhstan, Iraq, Iran, Afghanistan, Pakistan, India and China, wintering in Iran, Pakistan and northwest India) (BirdLife International 2001, 2013).

In the Indian subcontinent, the Marbled Duck is generally rare and local, with small numbers breeding and wintering in (chiefly southern) Pakistan, wintering in northwest India (with a few records from Assam and a report from Bangladesh) (Ali & Ripley 1987, Grimmett *et al.* 1998). Rahmani & Islam (2008) and Rahmani (2012) have listed all the records from India. In the Kashmir Valley, specimens

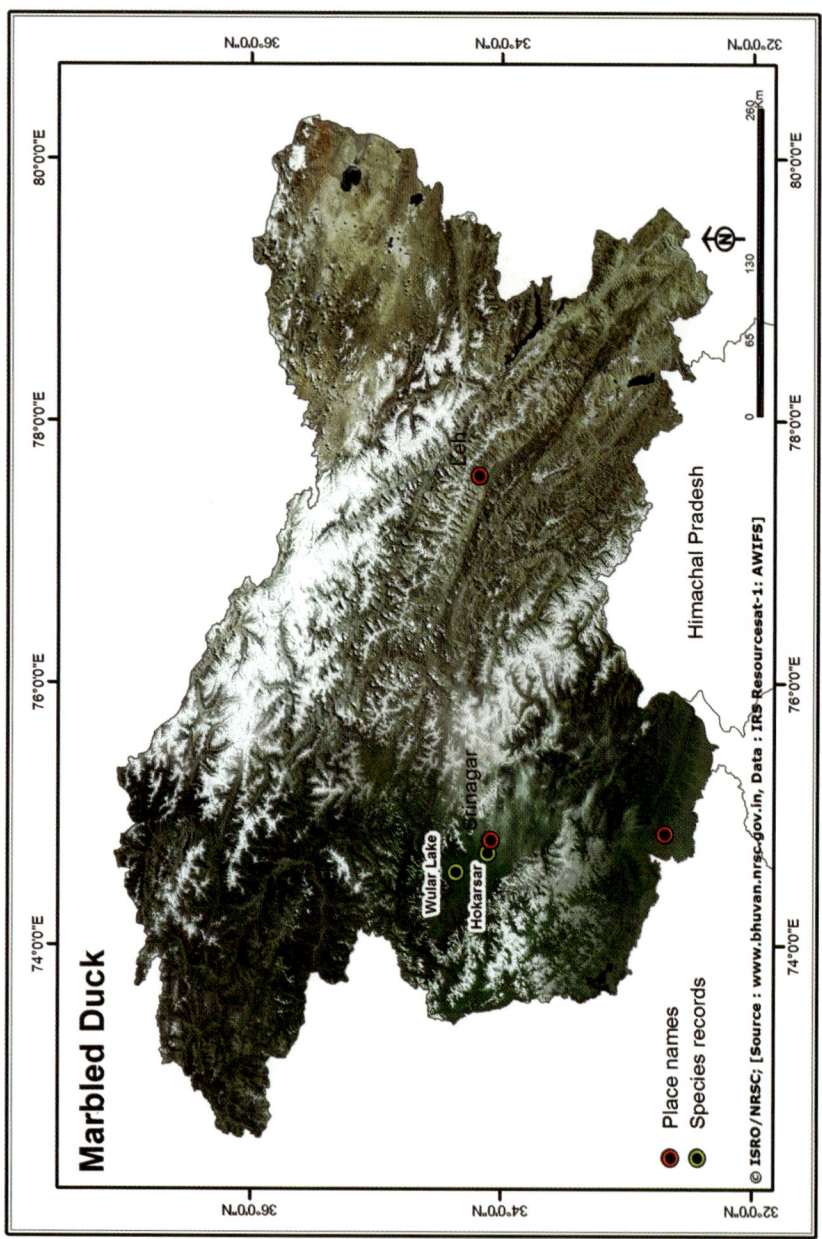

Marbled Duck

Wular Lake
Hokarsar
Srinagar
Leh

Himachal Pradesh

● Place names
○ Species records

© ISRO/NRSC; [Source : www.bhuvan.nrsc.gov.in, Data : IRS-Resourcesat-1: AWIFS]

Mirgund Jheel is still in fairly good condition despite some encroachment and pollution from surrounding households

have been collected from the **Wular Lake** in April 1923; **Mirgund Jheel**, near **Hokarsar**, in December 1935, and **Srinagar** in October 1913 (BirdLife International 2001). There is no recent sight record from the state (Bacha *pers. comm.* 2013).

Ecology: As it is a marginal species to the state, we are describing its ecology in brief. The Marbled Duck is found in jheels and shallow wetlands rich in emergent vegetation. In its breeding areas, nests are recorded from mid-April to late June and broods from mid-April to mid-September. The Marbled Duck is a shy, almost silent bird that usually lives in pairs or small parties (Green 2000, BirdLife International 2001). Large monospecific flocks are quite often formed in the post-breeding season and in winter; at dusk such flocks often fly from daytime roosts (with dense emergent vegetation) to more exposed, shallow feeding sites (BirdLife International 2001).

Threats: It suffers from the usual biotic pressures faced all over the world by wetlands: drainage for agricultural or industrial purposes, pollution, and excessive livestock grazing. As it breeds in tall dense reeds, reed-cutting, reed-burning, and grazing are a great threat during nesting.

Conservation measures underway: It is listed in Schedule IV of the Indian Wildlife (Protection) Act, 1972, and also in CMS Appendix I and II. It is one of the species to which the Agreement on the Conservation of African-Eurasian Migratory Waterbirds (AEWA) applies (BirdLife International 2013).

RECOMMENDATIONS

As it is a marginal species in Jammu & Kashmir with no recent record, we do not have any specific recommendation except that a thorough survey should be conducted in all the wetlands of the Kashmir Valley, and strict control on poaching should be maintained for all waterfowl.

Pallas's Fish-eagle
Haliaeetus leucoryphus (Pallas 1771)

RISHAD NAOROJI

BirdLife International (2013) justifies its inclusion in the Vulnerable category as this species has a small, declining population as a result of widespread loss, degradation, and disturbance of wetlands and breeding sites throughout its wide range.

Field Characters: A large (76–84 cm), dark brown eagle with pale golden brown head and neck, and blackish tail with a broad, white, central band. The band is particularly conspicuous in flight. Juvenile more uniformly dark, with all-dark tail, but in flight shows strongly patterned underwing, with whitish band across coverts and prominent whitish primary flashes. Female is like the male but larger.

The voice is a loud, guttural *kha-kha-kha-kha* or *gao-gao-gao-gao*, and it sometimes makes a high-pitched, excited yelping noise. The commonest call is a hoarse, guttural, continuous *kook-kook-kook*.

Distribution: Pallas's Fish-eagle ranges extensively from Kazakhstan, southern Russia, Tadjikistan, Turkmenistan (probably dispersing non-breeders), and Uzbekistan, east through Mongolia and China, south to northern India, Pakistan, Afghanistan, Nepal, Bhutan, Bangladesh, and Myanmar. The main breeding populations are believed to be in China, Mongolia, and the Indian subcontinent. It appears to have declined significantly during the 20th century in China, Pakistan, India, Nepal, and Bangladesh. The population is likely to be <10,000 mature individuals (BirdLife International 2013).

Pallas's Fish-eagle

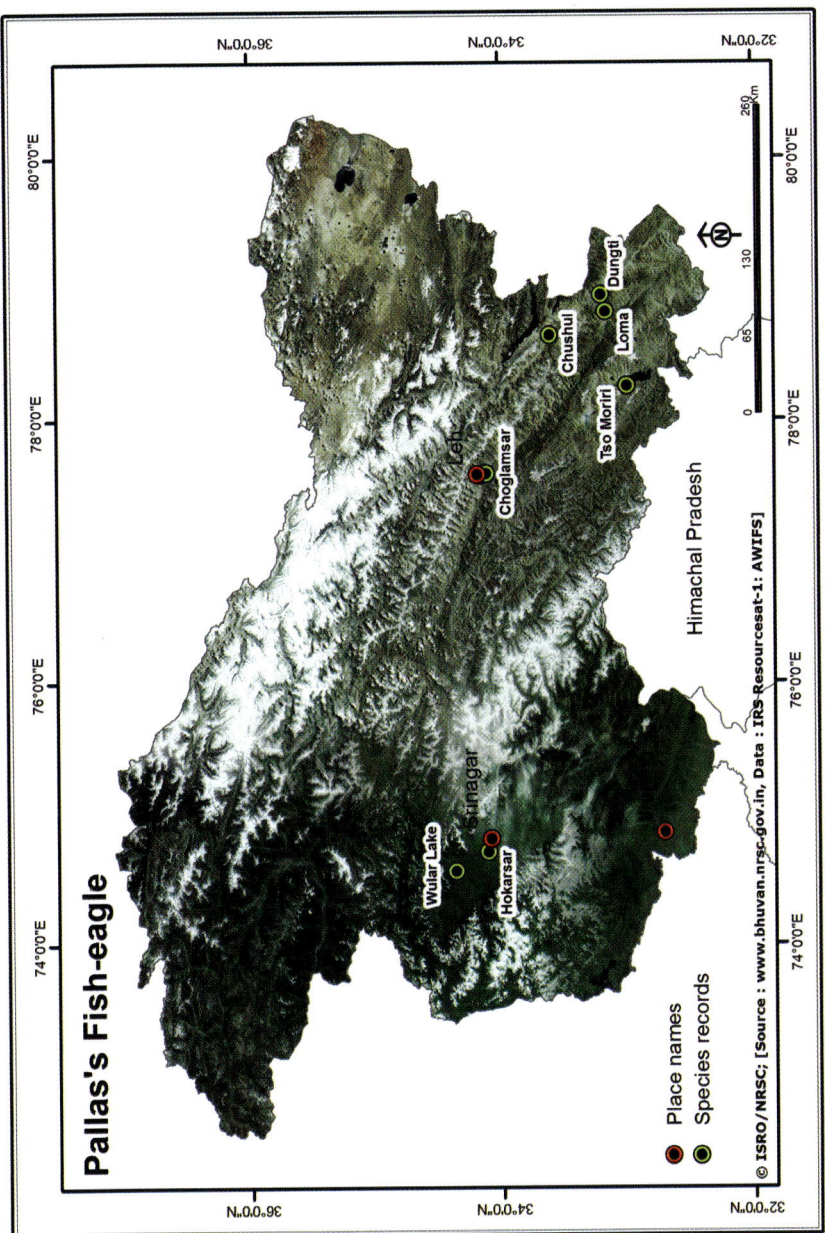

Place names ●

Species records ◉

© ISRO/NRSC; [Source : www.bhuvan.nrsc.gov.in, Data : IRS Resourcesat-1: AWIFS]

Pallas's Fish-eagle was sighted in Dungti area in May 2011, perhaps on migration

Rahmani (2012) has given recent and some old records from India. In Kashmir, it was historically reported from **Wular**, **Hokarsar**, and around the **Jhelum river**. Lawrence (1895) mentions that it is "pretty common in spring, summer, autumn and winter on the Jehlum and the country near the Wular lake." In 1889, he shot one bird at "Hajun" which "was seated on a high Chenar-tree" and whose "plumage indicated that of a young bird". There is no recent record of the bird from the Kashmir Valley or Jammu region.

Williams & Delany (1996) reported a spring migrant from Ladakh, where it is a scarce resident and passage migrant (Naoroji 2007). Pfister (2004) also reported it as a rare passage migrant in spring (May-June) through the lower depths of the valley of central Ladakh and the high altitude lakes of eastern Ladakh up to 4,600 msl. Pfister (2004) had single observations from the **Choglamsar/Leh-Tikse area**, **Tso Moriri**, and **Chushul**. Pankaj Chandan and Nisa Khatoon (*pers. comm.* 2012) sighted this bird in **Loma** (May 2009) and **Dungti** (May 2011) during avifaunal surveys of high altitude wetlands of Ladakh.

Ecology: Pallas's Fish-eagle is invariably found near water, mainly large wetlands, rivers, and jheels, from lowlands to 5,000 msl. It feeds mainly on fish, which are sometimes too large for it to lift, and small waterfowl. It is well known for pirating prey from Osprey, Marsh Harrier, and Brahminy Kite, continuously harassing them till the prey is relinquished. In heronries, it feeds on chicks of cormorants, Oriental Darter, egrets, ibises, and Asian Openbill, and is capable of

Threatened Birds of Jammu & Kashmir

killing birds as large as the Common Crane and Bar-headed Goose. In earlier literature, it was mentioned that Pallas's Fish-eagle takes a heavy toll of young Bar-headed Geese in the Ladakh lakes.

It generally nests on large trees near water. In India, the nesting season is from October to February. For details of its ecology and behaviour, see Naoroji (2007).

Birds breeding in northern climes are migratory and leave the area by October, and first breeders return by end-March. In north India, both resident and migratory birds are seen in winter, when the resident birds breed, while adult and immature birds arrive from Mongolia, China, and Central Asia. After breeding, the resident birds move from hot lowlands to cooler higher areas.

Threats: In India, the Pallas's Fish-eagle suffers the problems faced by almost all waterbirds. However, as it is at the apex of the food pyramid, it is the first to disappear due to degrading of wetlands, spread of Water Hyacinth *Eichhornia crassipes*, over-fishing, felling of large trees near wetlands, indiscriminate use of agricultural pesticides, and industrial runoff into the wetlands.

The extensive range of the species is deceptive, as Pallas's Fish-eagle is rare and isolated throughout its territory, and may not breed in large parts of it. Its overall range may be 1,000,000 sq. km, but the birds are scattered very sparsely throughout (Ferguson-Lees & Christie 2001).

Conservation measures underway: It is listed in Schedule IV of the Indian Wildlife (Protection) Act, 1972, and also CITES Appendix II and CMS Appendix II.

RECOMMENDATIONS

BirdLife International (2013) has given many suggestions for conservation action in the whole range of the species, while Rahmani (2012) has given India-specific recommendations. Here we give state-level recommendations:

(1) Promote rural education programmes concerned with wetland birds.
(2) Study its movement and dispersal through ringing/tagging and satellite telemetry.
(3) Pesticide level in prey species should be monitored, and if high, remedial measures taken. This is particularly valid for Kashmir Valley and Jammu region.
(4) Organic farming should be encouraged around important wetlands in the Kashmir Valley.

Greater Spotted Eagle
Aquila clanga Pallas 1811

DHRITIMAN MUKHERJEE

BirdLife International (2013) lists it as Vulnerable as it has a small population which appears to be declining owing to extensive habitat loss and persecution.

Field Characters: A very dark eagle of 62–74 cm, invariably found near large jheels and wetlands. On perch at close range, the best way to separate it from other Aquilas is by its round nostrils and the gape line stopping at the centre of the eye. This facial characteristic is consistently different in all Aquila eagles which are sometimes found in the same area in winter. On closer examination, an adult is dark brown (not black) with purplish or maroon sheen on the mantle, with slightly paler ventral side. It also has slightly paler flight feathers, with underwing coverts generally darker than flight feathers. Sexes alike, but female is larger. Juveniles paler, with back and wings sparsely spotted or streaked with buff or white. For age-related plumage differences, see Naoroji (2007). In gliding flight, it often depresses its "hands". It can be confused with other Aquila eagles, but

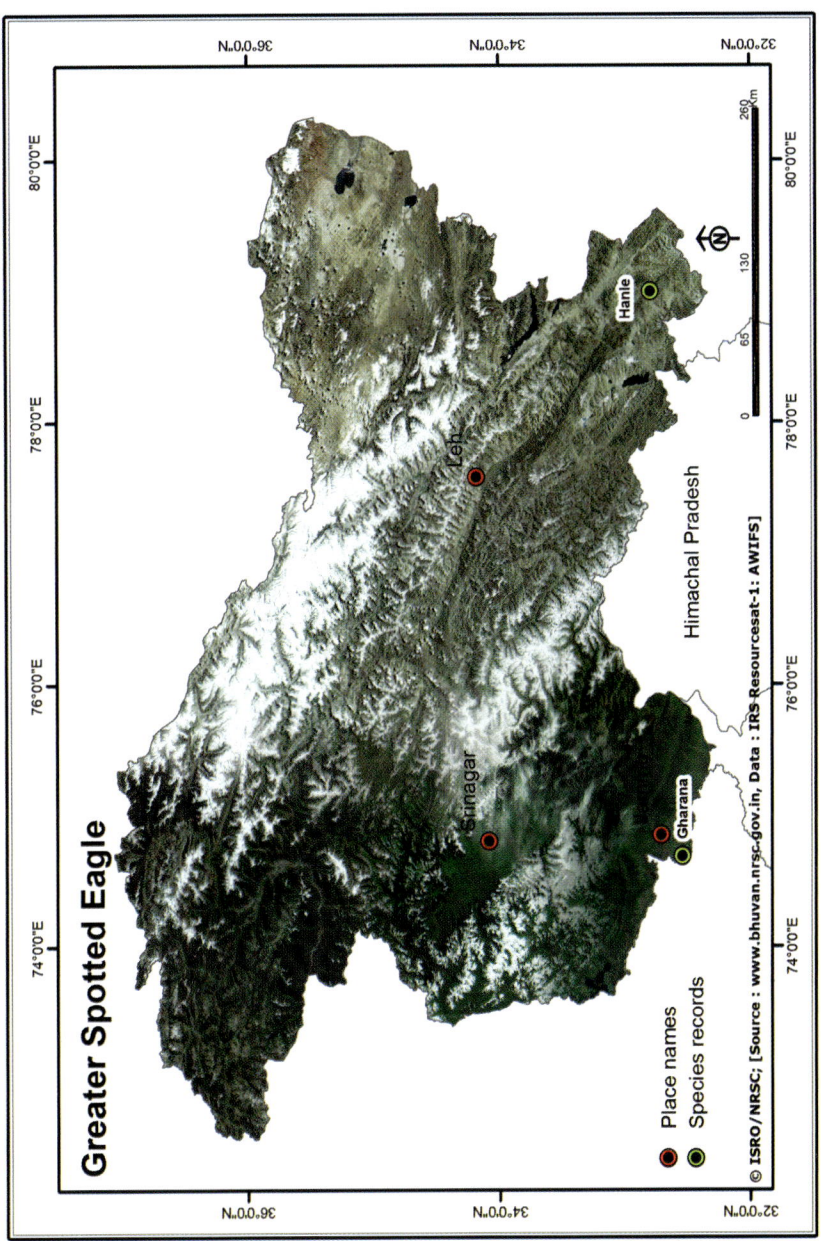

Greater Spotted Eagle

Place names
Species records

© ISRO/NRSC; [Source : www.bhuvan.nrsc-gov.in, Data : IRS-Resourcesat-1: AWiFS]

Himachal Pradesh

Leh

Hanle

Srinagar

Gharana

its habit of living near water sometimes makes it easy to identify. Pale morph called fulvescens (only in juvenile and immature: Rasmussen & Anderton 2005) is also seen although it is extremely rare.

Distribution: In India, the Greater Spotted Eagle winters widely. In Jammu & Kashmir, there are a few recent records. Pfister (2004) says "A rare vagrant on autumn passage migration through the high altitude plains of eastern Ladakh at altitudes up to 4300 m." He found a single record from Raar in the **Hanle** plains in early August. We do not have any confirmed record from the Kashmir Valley (Bacha *pers. comm.* 2013). In the Jammu region, one bird has been sighted in **Gharana** wetland (Intesar Suhail *pers. comm.* 2013).

Ecology: The Greater Spotted Eagle is invariably found near water where it sits and waits for hours for the right prey. It preys upon waterfowl, particularly the sick and injured, and chicks from heronries, as seen in Keoladeo NP. In some parts of the world it mainly feeds on terrestrial prey — for example, in wet grasslands, it feeds on amphibians and small mammals. It is also found on rubbish dumps. Its diet is highly variable. Basic information on its ecology and behaviour is given by Ferguson-Lees & Christie (2001) and Naoroji (2007).

Threats: In India, drainage and degradation of wetlands is the biggest threat to this species, and all waterbirds in general. BirdLife International (2013) has listed other causes such as shooting in Belarus, Russia, and Poland, the Mediterranean and Southeast Asia, accidental or deliberate poisoning, and hybridisation with the more numerous Lesser Spotted Eagle *Aquila pomarina*.

Conservation measures underway: Like all large birds of prey, the Greater Spotted Eagle is also protected under the Indian Wildlife (Protection) Act, 1972 and listed in Schedule I. It is listed in CITES Appendix II, and CMS Appendix I and II. It is found in many IBAs/PAs and other sites in India (Islam & Rahmani 2004).

RECOMMENDATIONS
(1) Start regular surveys to know its status, range, and population trends.
(2) Maintain, improve, and protect large wetlands through appropriate legislations and community conservation.
(3) Improve understanding of breeding habitat requirements and protect breeding areas, if found.
(4) Study its movement through satellite tracking (PTT).

Eastern Imperial Eagle
Aquila heliaca Savigny 1809

VISHWATEJ PAWAR

BirdLife International (2001, 2013) has listed the Eastern Imperial Eagle as Vulnerable, as it has a small global population, and is likely to undergo continuing decline, primarily as a result of habitat loss and degradation, adult mortality through persecution and collision with power lines, nest robbing and prey depletion.

Field Characters: The Eastern Imperial Eagle is a large Aquila, up to 84 cm, slightly smaller than the Golden Eagle *Aquila chrysaetos* (75–88 cm). It is stout-bodied with long and broad wings, longish tail and distinctly protruding head and neck (Grimmett *et al.* 1998). It is dark brown overall with white scapular markings and pale golden-cream nape, that clinches identification (in adult). Juvenile creamy-buff with dark streaks on underparts up to belly, with dark flight feathers (see picture above). For details of age-related plumage pattern, see Rasmussen & Anderton (2005) and Naoroji (2007).

Distribution: This eagle has a large range, breeding mainly in the Palaearctic from central Europe to the Russian Far East, and wintering in the African and

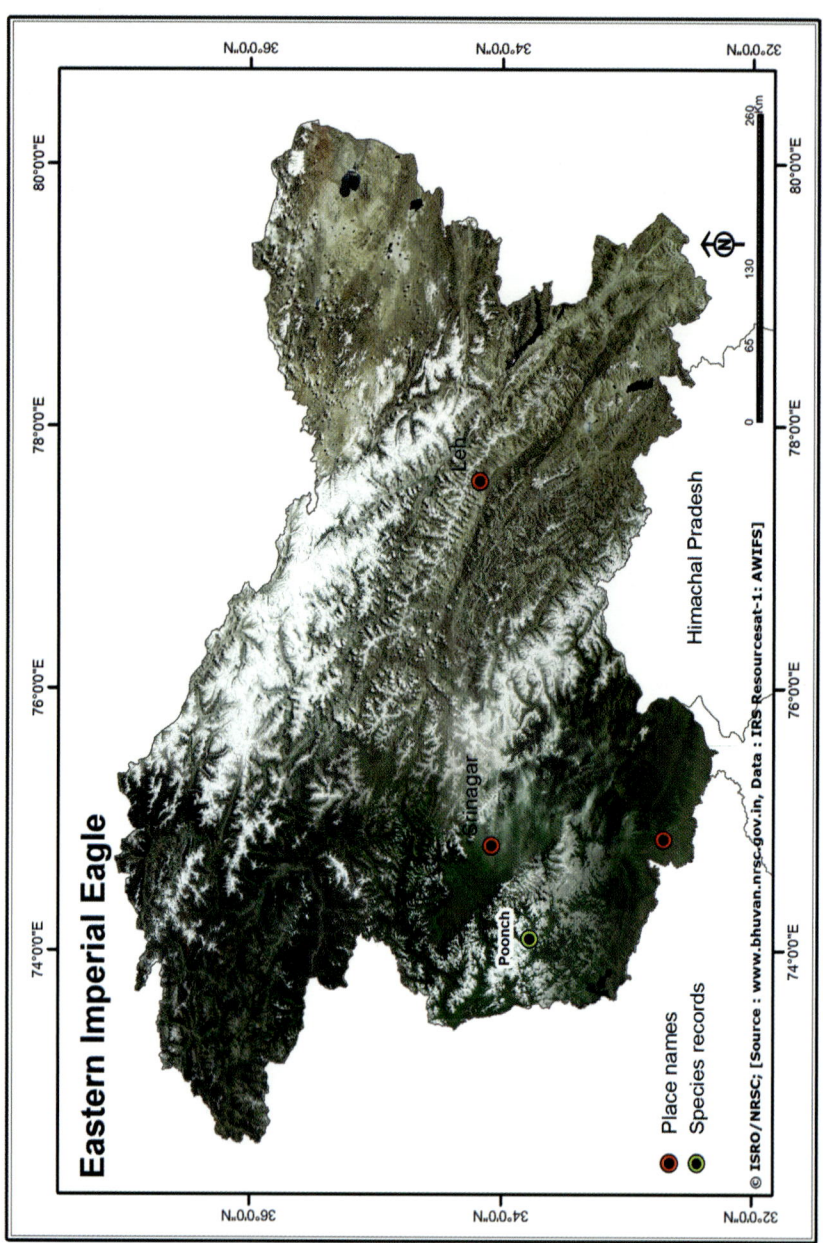

Eastern Imperial Eagle

Place names ●
Species records ●

© ISRO/NRSC; [Source : www.bhuvan.nrsc.gov.in, Data : IRS Resourcesat-1: AWIFS]

Himachal Pradesh

Leh

Srinagar

Poonch

Threatened Birds of Jammu & Kashmir

Oriental regions. In the Indian subcontinent, it is a winter visitor, particularly to northern India in open treeless country in the plains and deserts, mostly around wetlands. Its main wintering range extends from Pakistan, eastwards to Nepal, south through Kutch and Saurashtra to Mumbai, eastwards to central Maharashtra (Naoroji 2007).

In Jammu & Kashmir there are very few records. Pfister (2004) has not listed this species from Ladakh, and Mr. Bacha (*pers. comm.* 2013), a very experienced birdwatcher of Kashmir has not seen this bird in the Kashmir Valley. However, Singh (2013) photographed this bird on December 5, 2011 at about 11.20 am, 5 to 7 km downstream of **Poonch** town, slightly above the Poonch river.

Ecology: Ali & Ripley (1987) have succinctly described its behaviour in its winter quarters: "A heavy sluggish eagle, normally seen perched for hours on end on a stump or tree-top, or on the bare ground, in open semi-arid or flat featureless country such as at the edge of the Rann of Kutch. Obtains its food by pouncing on any small animal that may show itself in its vicinity, but mostly by pirating — chasing other hawks and eagles — and forcing them to surrender what they have hunted." It also eats carrion and so can be seen around carcass dumps and slaughterhouses (e.g., Jor Beer near Bikaner). Naoroji (2007) found that some individuals use the same area and perches year after year, showing site fidelity. He also found that among Aquilas, this species is most dominant in piracy.

Threats: Habitat alterations associated with agricultural expansion threaten historical and potential breeding sites in former range countries. Human disturbance of breeding sites, nest robbing, and illegal trade, shooting, poisoning, and electrocution by power lines are some more threats. Hunting, poisoning, prey depletion and other mortality factors are also likely to pose threats along migration routes and in wintering areas (BirdLife International 2013).

Conservation measures underway: Like all large birds of prey, the Eastern Imperial Eagle is listed in Schedule I of the Indian Wildlife (Protection) Act, 1972. There is no specific action plan in India. There are many protected areas and IBAs in India where it is found.

RECOMMENDATIONS

As it is marginal to the state, we do not have any specific recommendations for this species. General recommendations for all raptors, i.e., regular surveys to know their status, range, and population trend, also apply to this eagle.

Sarus Crane
Grus antigone (Linnaeus 1758)

ASAD R. RAHMANI

The justification for including Sarus Crane in the Vulnerable category is that it has suffered a rapid population decline, which is projected to continue, as a result of widespread reduction in the extent and quality of its wetland habitats, and mortality from poisoning and pollutants (BirdLife International 2013).

Field Characters: Sarus Crane is the tallest flying bird in the world. It stands 152–156 cm tall. The adults are grey overall, with whiter mid-neck and tertials, mostly naked red head and upper neck, blackish primaries, mostly grey secondaries, and reddish legs that are bright during the breeding season and pale red otherwise. Females are supposed to be slightly smaller, but sometimes this difference is imperceptible. Subspecies *Grus antigone sharpii* is a more uniform, darker grey. The bare red skin of the adult head and neck is also brighter during the breeding season. This skin is rough and covered by papillae, and a narrow area around and behind the head is covered with black bristly feathers. Juveniles have a feathered, buffish head and upper neck, and duller plumage with brownish feather fringes.

Sarus Crane

Place names

Species records

© **ISRO/NRSC;** [Source : www.bhuvan.nrsc-gov.in, Data : IRS-Resourcesat-1: AWIFS]

Srinagar

Leh

Kathua

Himachal Pradesh

In Jammu & Kashmir, the Sarus Crane is found only in the Jammu region but all the wetlands and jheels that it inhabits are intensively used by human beings

Distribution: The Sarus Crane has three disjunct populations – in the Indian subcontinent, Southeast Asia, and northern Australia. In India, it is found in many states (see Rahmani [2012] for historical and recent distribution).

There is now enough evidence of Sarus Crane's occurrence in Jammu region. It was reported from **Kathua** district, at Kishanpur **Gharana** Wetland Reserve (Sahi 1993). Sundar *et al*. (2000), during a survey of Sarus in India, counted 24 birds from three sites in Kathua district. Two adult birds were seized from poachers by the Department of Wildlife (Protection) Jammu & Kashmir, from Kathua district a few years ago.

In the Kashmir Valley, the bird has been reported on two rare occasions, undated (Ward 1906–1908). Since there is no record of this species in the Kashmir Valley before or after Ward's unconfirmed record, his record is questionable.

Ecology: The Sarus uses open, wet and dry grasslands, agricultural fields, marshes, and jheels for foraging, roosting, and nesting (Sundar *et al*. 2000; Sundar & Choudhury 2003). Wetlands, even those of very small size and close to roads and human habitation, are the preferred habitat for nest construction (Sundar 2009). For foraging, Sarus usually uses crop fields to a lesser extent, and prefers feeding in wetlands. It is omnivorous, and feeds on a variety of roots and tubers, as well as invertebrates and amphibians. Rahmani (2012) has described the latest researches on the Sarus Crane.

Threats: In India, Sarus is considered a sacred bird, so hunting is not the main problem, it is habitat destruction and habitat alteration which are taking their toll on its population. Wetlands are under tremendous pressure from human use, drainage and conversion to agriculture, housing colonies, and even construction of highways. Recently, a hazard has emerged in the form of pesticide poisoning (Muralidharan 1992, Kaur & Nair 2008). Collision with power lines may be a significant threat in parts of its range, with observations from India suggesting that 2.5–20% of some populations are affected by such stochastic occurrences (Sundar *et al.* 2000, Sundar & Choudhury 2001, Rana & Prakash 2004), annual mortality of mostly non-breeding birds equalling almost 1% of the total population in some areas (Sundar & Choudhury 2006). The other major reason for the decline in numbers of the Sarus Crane is egg mortality (Mukherjee *et al.* 2001, Kaur & Choudhury 2003, Sundar 2009). Eggs are preyed upon largely by crows (Ramachandran & Vijayan 1994, Sundar 2009).

According to Sundar (*pers. comm.* 2010), chick predation by dogs and egg predation by corvids is increasing, as their populations increase following the decline of vultures in the Indian subcontinent.

Conservation measures underway: The Sarus Crane is listed in Schedule IV of the Wildlife (Protection) Act, 1972, and in CITES Appendix II and CMS Appendix II. It is the State Bird of Uttar Pradesh, where nearly 50% of India's Sarus population is present. Reports of Sarus Crane in Jammu & Kashmir are mainly from unprotected areas.

RECOMMENDATIONS

(1) Conservation of all wetlands in Jammu region should be accorded the highest priority. Restoration of wetlands with local participation is most important in this state, but is also required in other states.

(2) Detailed research is required on local and seasonal movements of Sarus Crane through colour banding and satellite tracking.

Black-necked Crane
Grus nigricollis Przevalski 1876

DHRITIMAN MUKHERJEE

Black-necked Crane is the last of the world's cranes to be discovered by the scientific community, when it was first sighted by the Russian naturalist Przevalski near Lake Koko Nor in northeast Tibet, in 1876. In Jammu & Kashmir, it was first reported in Ladakh in 1919 at Tso Kar (Ludlow 1920).

The Black-necked Crane is listed by BirdLife International (2013) as Vulnerable. Though it is believed that the numbers are probably increasing, it still faces numerous threats from loss and degradation of wetlands, and changing agricultural practices in both its breeding and wintering grounds, and dedicated monitoring is needed. Its global population is estimated to be 6,600 mature individuals. In India, although the population is very small, field studies show an increase in numbers. Changing agricultural practices and climate change are long-term threats to the species.

Field Characters: A tall bird, with a height of *c*. 139 cm and wingspan of *c*. 235 cm, the adult weighs *c*. 6 kg. Both the sexes are almost the same size but the male is slightly larger. The long upper neck, head, primary and secondary flight feathers are completely black, and the body plumage ashy grey, which becomes whitish on the underparts. The tail is black, and uppertail-coverts are greyish. The black primaries and innermost elongated secondaries are easily visible in flight. The smaller wing-coverts are pale grey. Iris is yellow. A conspicuous red crown adorns the head. The bill is greenish, with a yellowish tip. Legs and toes are black. Hairlike feathers sparsely cover the lores.

Map of locations having presence of Black-necked Crane in Changthang, Ladakh.

Fifteen to sixteen pairs of Black-necked Crane are known to breed in Ladakh

The chicks are covered with brownish down, and have a flesh-red bill, which becomes whitish at the tip. The feet are reddish, with a touch of grey. By the time the chick is 20 days old, the colour of the head and tail becomes darker. At four weeks, the toes are greyish brown, the primary quills begin to emerge and the top of the head becomes pale yellow. The juvenile has a brownish head and neck, with plumage slightly paler than that of the adult. Juveniles at 70 to 90 days have yellow-brown feathers on the crown, and grey abdomen. The primaries, secondaries and the feathers of the back are greyish yellow. By eight months, the iris is yellow brown, and one-third of the neck is greyish black, with some yellowish brown feathers remaining on the back. By the time it reaches one year of age, the juvenile resembles the adult.

This species which inhabits the high altitude wetlands of the Tibetan Plateau can survive temperatures down to -30 °C.

Distribution: The Black-necked Crane breeds on the Qinghai-Tibetan Plateau, with a small population of about 120 individuals in Ladakh in India. It winters in Bhutan (500 individuals), and Arunachal Pradesh (10 individuals). Its range covers 28–38° N and 78–104° E, stretching from the Altun and Kunlun mountain ranges east to the Qilian and Wumeng mountain ranges, and south to Himalaya (Chandan *et al.* 2005).

The population of Black-necked Crane has increased, perhaps due to better protection, but it is not known whether these figures represent short-

Thanks to good protection by the Wildlife Department, army and by civilians, there has been steady increase in the number of Black-necked Crane in Ladakh

term population peaks, or long-term trends (BirdLife International 2013). The main reason for the increasing numbers recorded is that more coordinated surveys have been conducted, both in the wintering and the breeding grounds.

The main wintering grounds of the Black-necked Crane are Tibet and Bhutan (Chacko 1992a, 1993a), and small numbers also winter in India, mainly Arunachal Pradesh. For example, 20–40 birds were recorded in most years until the 1940s, and 27 birds in 1946 (Betts 1954) in Apatani Valley in Lower Subansiri. However, due to development of large towns Ziro and Hapoli in the valley and resultant disturbances, the cranes have stopped coming since the 1960s, except for a few stragglers in the 1970s. After their disappearance from the Apatani Valley, it was believed that they do not winter in India anymore, until a new site was found in Sangti Valley in West Kameng district by Gole (1996). Choudhury (1996) reported stray records from places such as Nyapin and Palin in Lower Subansiri district and an unconfirmed report from Namdapha NP. Even now, every year a small number (one to six cranes) visit the Sangti Valley. In the winter of 1999–2000, three were photographed by A.U. Choudhury (*The Twilight* 2 [2&3]). The cranes did not visit the Sangti Valley in 1995 and 1996. In 1999, Choudhury found evidence of wintering of Black-necked Crane in Nyamjang Chu (chu = river) at Zemithang in Tawang district. Zemithang is 95 km from Tawang and 250 km from the Sangti Valley by road, but only 20 km and 65 km respectively as the crow flies. The local name of the crane is *thungthung karmo*. According to Buddhist lamas in the local gompas, the cranes have been wintering in small numbers (maximum recorded seven) in Zemithang since time immemorial.

PANKAJ CHANDAN

It is unusual to see two juveniles in one family. Generally one chick is raised in a year

Month/Year	Number of Black-necked Crane	Breeding Pairs	No. of Wetlands covered	Reference
Jun 1919	3	1	2	Ludlow (1920)
Jun 1924	11	4	7	Osmaston (1925)
May-Jun 1926	10	5	8	Meinertzhagen (1927)
Jun 1976	5	2	4	Hussain (1976)
Jul 1978	12	1	10	Gole (1981)
May-Jun 1980	14	3	10	Gole (1983)
Jun 1982	13	3	9	Nurbu (1983)
Jun 1983	7	2	6	Hussain (1985)
Aug-Oct 1986	16	2	8	Narayan et al. (1987)
Jul-Nov 1987	9	1	5	Akhtar (1989)
Sep-Oct 1992	17	4	14	Chacko (1992a)
May-Sep 1995	22	5	18	Chacko (1995)
May-Aug 1996	25	12	18	Chacko (1996)
Jun-Sep 1997	38	12	18	Pfister (1998)
Apr-Dec 2002	59	15	22	WWF-India
Apr-Nov 2003	60	16	22	WWF-India
Apr-Nov 2004	64	15	22	WWF-India
Apr-Nov 2005	58	15	22	WWF-India
Apr-Nov 2006	59	15	22	WWF-India
Apr-Nov 2007	58	16	22	WWF-India
Apr-Nov 2008	81	15	22	WWF-India
Apr-Nov 2009	65	15	22	WWF-India
Apr-Nov 2010	73	11	24	WWF-India
Apr-Nov 2011	79	14	24	WWF-India
Apr-Nov 2012	139	14	24	WWF-India
Apr-Nov 2013	95	14	22	WWF-India

Black-necked Crane in Ladakh over the years (source: Pankaj Chandan)

Population: The total population of Black-necked Crane is estimated to be c. 10,900 (Birdlife International 2013). The Ruoergai marshes, located on the north-eastern edge of Qinghai-Tibet Plateau, have the largest breeding population, estimated at c. 2500.

Three wintering sub-populations have been recorded in the wintering grounds of Black-necked Crane. The eastern population of the species at Dashanbao, Caohai, and Huize (northeastern Yunnan and northwestern Guizhou provinces)

Four pairs of Black-necked Crane are known to breed in the Hanle Marshes

is estimated to be about 2,469 birds. The central population at Napahai Provincial Nature Reserve (northwestern Yunnan) has a population of 270. The western population is in south-central Tibet, Bhutan, and Arunachal Pradesh. The Yarlung Tsangpo Black-necked Crane National Nature Reserve hosts the majority of wintering Black-necked Crane in Tibet. This reserve covers almost all wintering areas in Tibet.

There are confirmed reports of the presence of Black-necked Crane in the remote valleys towards the north and west of Apatani Valley (Choudhury 2002). A survey by a WWF-India team reported nine individuals from Sangti Valley in Arunachal Pradesh in January 2006.

During the authors' survey in July 2007, 26 Black-necked Crane were recorded in the Changthang region, including seven breeding pairs in and around Hanle Marshes, and two breeding pairs in and around Chushul Marshes.

WWF's High Altitude Wetlands project team in Ladakh reported 128 adults and 11 chicks from Ladakh in 2012 (Pankaj Chandan *pers. comm.* 2012). This is the highest number of cranes ever recorded in Ladakh.

Ecology: Among the 15 crane species in the world, the Black-necked is the only crane found in the alpine regions. Its breeding habitat ranges from 2,600 to 4,800 msl and wintering grounds from 2,000 to 3,800 msl. The high altitude marshes and lakes of the Tibetan Plateau (Tibet, Qinghai, Xinjiang, Gansu), and eastern Ladakh (India) are the known breeding grounds of Black-necked Crane. The species is a winter visitor in Yunnan and Guizhou (China), and Bhutan (Pfister 1998). Shallow lakes, river banks, and small ponds are their preferred roosts.

The breeding habitats of the Black-necked Crane in Ladakh can be broadly classified into two categories, Lacustrine Marshes and Riverine Marshes:

Lacustrine Marshes are the most common wetland type found in eastern Ladakh where the bird breeds. These wetlands with small mounds provide excellent breeding habitats for the birds. The lacustrine marshes range from 4,000 to 4,800 msl. Most of the wetlands are freshwater or brackish, but some wetlands like Tso Kar (at 4,600 msl) are saline. From December to March, these wetlands remain frozen. Annual rainfall is usually less than 100 mm, glaciers and winter snowmelt is the major source of water to these wetlands.

Riverine Marshes in India are found in the floodplains of the Indus river. The main examples of such habitats in the Black-necked Cranes' range are Fukche, Dungti, and Staklung. These wetlands provide abundant food in the form of tubers, snails, and fish. The common vegetation species found in the marshes where these cranes breed are *Equisetum ramosissimum*, *Lepidium apetalum*, *Pedicularis longiflora*, *Utricularia minor*, *Polygonum nummularifolium*, and *Potentilla anserina*. In Ladakh, Pfister (1998) reported that chicks predominantly feed on insects, other invertebrates, and plant tubers. The feeding pattern of the birds during the breeding season depends on food availability. In some wetlands like Hanle, where there is an abundance of fish, the birds are quite often seen feeding on fish. On the other hand, in some wetlands like Yaya Tso, where fish are not abundant, they are seen to feed on tubers and aquatic insects most of the time. By nesting at high altitudes, the Black-necked Crane has relatively short migration routes among migratory crane species, with the longest ~700 km (Qian *et al.* 2009), and the shortest 200 km or less (Rinchen Wangmo 2007 RSPN Bhutan, *pers. comm.* to Pankaj Chandan). The cranes migrate along a north-south route over the Tibetan plateau.

Pairs show monogamy and site fidelity, coming back to the same marsh year after year. The female lays two eggs, but generally only one chick is raised, though there are many records in Ladakh of both chicks being reared successfully. Both sexes help in incubation and chick rearing. Detailed studies on the breeding biology were conducted by Pfister (1998), and earlier by Chacko (1992b). Studies on the ecology and biology of this species are being conducted by Pankaj Chandan of World Wide Fund for Nature, India.

Threats: BirdLife International (2001, 2013) has given in detail the various threats to this species in its main wintering and breeding areas in Tibet. Climate change and drying up or shrinkage of some wetlands due to global warming is now an increasing long-term threat.

In Ladakh in India, the major threat to the breeding population is increasing

PANKAJ CHANDAN

Ill-conceived plantation schemes of the Government in high altitude wetlands such has Hanle (above) have altered Black-necked Crane habitat (below)

KHURSHEED AHMAD

tourism and related infrastructure development. Another major threat to this population is the presence of a large number of dogs close to the nesting sites, who eat the eggs and chicks. Tremendous grazing pressure near the nesting areas is another major threat in Ladakh (Chacko 1993b, Pankaj Chandan *pers. comm.* 2010).

Major Threats in Ladakh

1. Loss of habitat: One of the major threats to successful breeding of Black-necked Crane in Ladakh is loss of habitat, e.g., Hanle Marshes, which are dying due to increased human pressure.

2. Feral dogs: A large number of feral dogs are a major threat to the breeding population of Black-necked Crane in Ladakh. In addition, dogs kept as pets by army personnel and by the local shepherds also damage the nests and eggs.

3. Human pressure: Changthang, the area where the bird breeds, has about 41 villages and hamlets with a total population of 12,000 settled and nomadic Changpas; 2000 Tibetan refugees who crossed the border during the early 1960s and have remained in Indian territory ever since.

4. Increase in livestock: Livestock yields 90% of the income of the local people. The increase in human population over the years has resulted in a corresponding increase in livestock population, leading to destruction of the wetlands.

5. Developmental activities: A potential threat to the successful breeding of Black-necked Crane in Ladakh is the unplanned and unregulated developmental activities that do not take into account the fragility of the area. An example is the raising of electric poles all across Hanle plains, quite close to the nesting sites of Black-necked Crane.

6. Tourism: The sudden influx of large numbers of tourists, especially at Tso Kar, Tso Moriri, Loma, and Startsapuk Tso, is a major threat to Black-necked Crane. This is a serious problem, as the peak period of breeding activity of the cranes coincides with the peak tourist season. Sometimes, just to get good photographs, tourists disturb the birds even during incubation.

7. Religious tourism: During summer some monasteries organize religious festivals that attract large numbers of pilgrims and tourists. For example, at Hanle a major festival is organized annually, when local people camp quite close to the nesting site of the Black-necked Crane.

8. Climate change: High altitude habitats where this crane breeds have experienced some of the most dramatic warming due to climate change. Black-necked Crane appears to benefit in the short term due to lowered mortality, with milder weather and increasing melt from the glaciers. But in the long run, when the glaciers have decreased or disappeared altogether, the lakes and marshes will shrink, resulting in loss of breeding habitat.

Conservation measures underway: Black-necked Crane is protected under Schedule I of the Wildlife (Protection) Act, 1972, so hunting and shooting it is totally prohibited. It is listed in CITES Appendix I and II and CMS Appendix I and II. It is legally protected in all range countries (Tibet, India, and Bhutan). Major breeding and wintering areas are protected in Tibet. There have been conservation and development programmes in local communities at the important sites of Caohai and Dashanbao. The Indian breeding population in Changthang Cold Desert Wildlife Sanctuary is protected by traditional Buddhist sentiment, and now by the armed forces. It is the State Bird of Jammu & Kashmir, and the emblem of the army battalion in Leh, Ladakh. The Black-necked Crane is

Black-necked Crane is a mascot of some army camps

one of the most sacred creatures for Buddhists, and is pictured alongside many of their deities in the monasteries of the region.

Status surveys have been conducted in India (e.g., Chandan *et al.* 2005) and Bhutan, and a WWF-India ecological study is ongoing in Ladakh by Pankaj Chandan.

RECOMMENDATIONS

BirdLife International (2013) has recommendations for the whole range of the Black-necked Crane, while Rahmani (2012) has given recommendations for India. In this book, we give recommendations for Ladakh:

(1) Tourism infrastructure near the breeding and wintering areas should be avoided.

(2) Approach to the nest for photographs, especially by tourists and casual visitors like the families of armed forces, are a threat to the breeding population in Ladakh. This should be strictly prohibited.

(3) Dogs in the breeding areas should be strictly controlled.

(4) Advertising and sign boards near the nesting sites should be avoided to prevent attracting the attention of tourists and casual visitors.

(5) There should be no plantation in the wetland areas where the bird breeds.

(6) Conversion of wetlands into agricultural lands should be avoided.

(7) Coordination between various NGOs, developmental agencies, and government departments responsible for conservation, should be put in place.

(8) Grazing livestock near nesting sites during summer should be avoided, but can be permitted during winter. This can be tried experimentally in collaboration with local communities.

Pale-backed or Yellow-eyed Pigeon
Columba eversmanni Bonaparte 1856

VISHWATEJ PAWAR

BirdLife International (2013) considers the Pale-backed Pigeon Vulnerable as it has declined rapidly in the past, probably as a result of changing agricultural practices and hunting in its wintering grounds, and possibly habitat loss in its breeding grounds.

Field Characters: A smaller and paler version of the Blue Rock Pigeon *Columba livia*, mostly grey but with brownish cast to upperparts. Yellow eyes and eye-ring are characteristic and lend the bird its popular name. It has whitish lower back, rump, and underwing, and diffused, dark tail with a terminal band. Three small, narrow black bars across secondaries are visible while sitting and in flight. Juvenile has brownish-tinged eyes and lacks gloss in plumage. Crown, hindneck, and breast tinged with vinous or lilac, and sides of lower neck metallic green (but not as much as in the Blue Rock Pigeon or Hill Pigeon *Columba rupestris*). Sexes are alike.

Distribution: The Pale-backed Pigeon is a winter migrant to northwest India in the plains, orchards, lightly wooded country, and cultivation in Jammu, Punjab, Haryana, Delhi, Uttar Pradesh, Rajasthan, and is reported from as far east as Bihar. It has been specifically recorded from Jammu & Kashmir, Ludhiana and

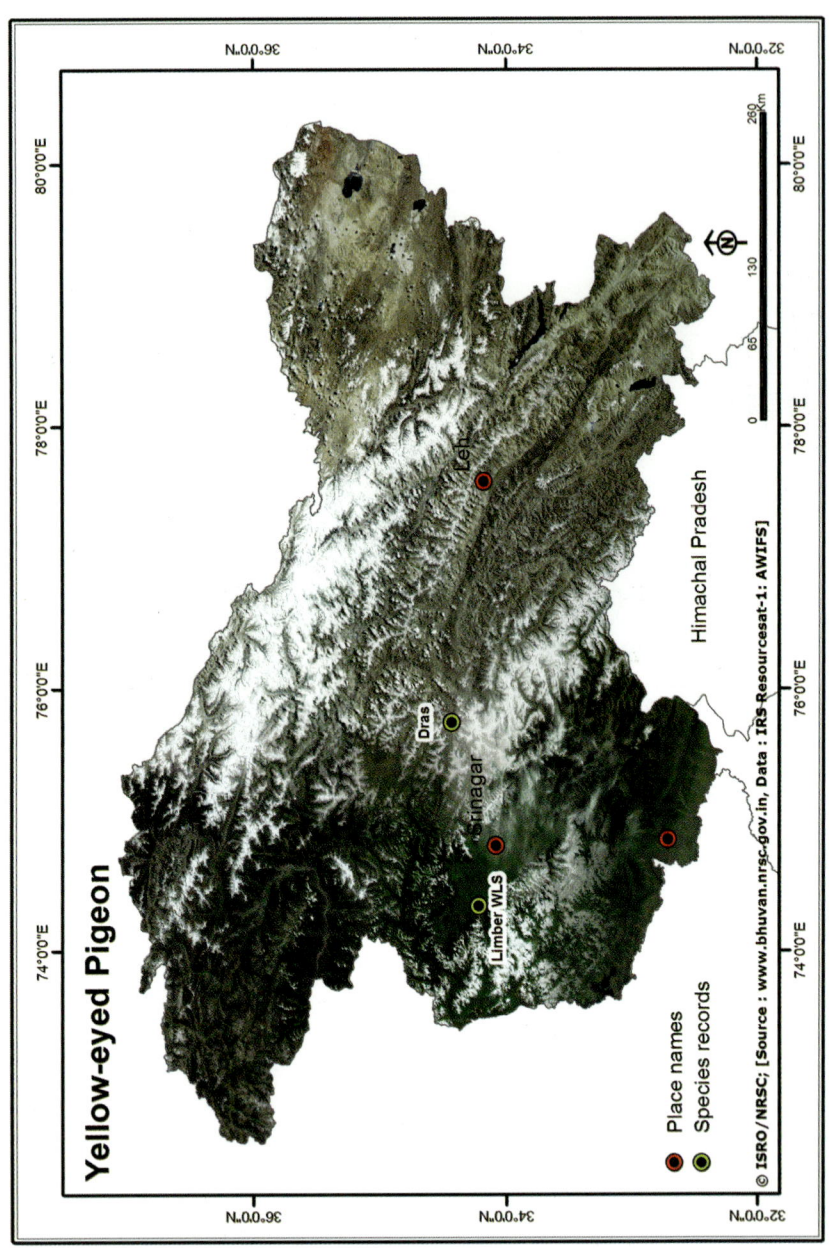

Yellow-eyed Pigeon

Place names
Species records

© ISRO/NRSC; [Source : www.bhuvan.nrsc.gov.in, Data : IRS-Resourcesat-1: AWIFS]

Leh

Dras

Srinagar

Limber WLS

Himachal Pradesh

Pale-backed Pigeon was earlier reported from Dras (above), Limber, and Chagra areas. It is perhaps still found in these areas during migration. More surveys are required

Ambala districts of Punjab, Lucknow and Gorakhpur districts of Uttar Pradesh, Darbhanga district in Bihar, and north Madhya Pradesh (Ali & Ripley 1987). Possibly they come to India in larger numbers during severe winters.

It is found in the desert and settled regions in southern Kazakhstan, Uzbekistan, Turkmenistan, Tadjikistan, Kyrgyzstan, Afghanistan, northeast Iran, and extreme northwest China. Its status and distribution within this range are poorly known (BirdLife International 2001). During the 19th and early 20th centuries, it was recorded in huge flocks from its wintering grounds, particularly in Punjab, India. However, it has declined rapidly, from wintering flocks numbering thousands of birds to flocks of tens or a few hundred birds, with occasional larger counts. Whether it continues to decline is unclear (BirdLife International 2013). Its global population could now be as low as 10,000.

Rahmani (2012) has given all recent records from India. Historically, it has been recorded from **Dras**, **Limber** WLS, and **Chagra** in Jammu & Kashmir but we do not have any recent record, perhaps mainly because there are very few good birdwatchers in the area.

Ecology: In its wintering quarters, it is seen in flocks, often with the Blue Rock Pigeon. Earlier, it used to be seen in thousands but now flocks of 10–15 are found, often less. However, in some areas, for example, Jor Beed near Bikaner, and Tal Chhaper in Churu district, both in Rajasthan, sometimes flocks consisting of hundreds of birds are found. Due to its similarity with Blue Rock Pigeon, it is often overlooked. Moreover, shooting of pigeons is prohibited, so confirmed specimen records as obtained earlier are hard to come by. Not many birdwatchers are present in the areas where it winters in India.

Like its other cousins, it nests in holes in trees, buildings, cliffs, and earth banks in semi-arid and desert areas, around human settlement and, at least in Kazakhstan, in woodland (del Hoyo *et al*. 1997). Breeding areas nearest to India are north and west Afghanistan, in treeless riverside cliffs to about 1500 msl (Rasmussen & Anderton 2005). It lays two eggs, and incubation and chick rearing are done by both parents. It feeds on grass seeds, crop seeds, ripening mulberries, and other small fruits.

Most populations are long-distance migrants, but the population in northwest China shows altitudinal movements (del Hoyo *et al*. 1997).

Threats: Hunting in both its breeding and wintering grounds has been the primary cause of its decline and continues to be a major threat in China. In India, intensification of cultivation and a change from the large-scale cultivation of pulses and mustard to wheat and rice has reduced the quality of habitat in its key wintering areas. Destruction of Poplar *Populus* woodland is believed to have had a major impact on the breeding population in eastern Kazakhstan (BirdLife International 2001, 2013). It is accidentally trapped by Blue Rock Pigeon trappers (Bhargava 2001).

Conservation measures underway: A protected species under Schedule IV of the Indian Wildlife (Protection) Act, 1972 in India, it is found in some protected areas such as Harike Lake Bird Sanctuary in Punjab, and Tal Chhapar in Rajasthan.

RECOMMENDATIONS

As very little is known in India about this species, we have the following recommendations:
(1) Thorough survey and monitoring during winter in its historical range in Jammu & Kashmir.
(2) Strict control on hunting of pigeons, particularly in winter when this bird is likely to be shot.
(3) Control on trapping of wild pigeons, particularly in winter.
(4) Conservation education to increase awareness and interest in this species.

Kashmir Flycatcher
Ficedula subrubra (Hartert & Steinbacher 1934)

CLEMENT FRANCIS

The Kashmir Flycatcher has a small, perhaps declining population, and a fragmented breeding range, as a result of the destruction of temperate, mixed deciduous forests. It therefore qualifies as Vulnerable (BirdLife International 2001, 2013).

Field Characters: The Kashmir Flycatcher is a small bird (13 cm), roughly equal to its cousin, the Red-breasted Flycatcher *Ficedula parva*. The male has a chestnut-rufous throat, breast, and flanks, with characteristic black malar streak extending to the upper breast-patch. In the Red-breasted, the red or rufuous patch is limited to the throat and top of the breast. Female of the Kashmir Flycatcher is very similar to female of Red-breasted, except that the Kashmir Flycatcher female has a rufous wash on chin and breast. Females and first winter birds have dark base to bill and paler, slightly browner upperparts. For more details, see Rasmussen & Anderton (2005).

Distribution: The Kashmir Flycatcher is endemic to the Indian subcontinent, breeding in the northwest Himalaya, the Neelum Valley and Kazinag Range in Pakistan (Roberts 1992), and the Pir Panjal Range in India. It occurs between 1,800–2,700 msl in its summer breeding areas and more or less at the same heights in winter in south India and Sri Lanka. It has a very restricted breeding range in northern India and in some parts of Pakistan.

Kashmir Flycatcher

Place names ●
Species records ●

© ISRO/NRSC; [Source : www.bhuvan.nrsc-gov.in, Data : IRS-Resourcesat-1: AWiFS]

Leh

Srinagar
Dachigam
Tral
Hirpora
Gulmarg
Dehra Gali

Himachal Pradesh

Zarri & Rahmani (2004) and Rahmani (2012) have given the latest wintering records of this bird from the Western Ghats and Sri Lanka. Here we describe summer (breeding) records from Jammu & Kashmir. In the book *Important Bird Areas in India* by Islam & Rahmani (2004), it has been reported from **Dachigam** NP, **Dehra Gali** Forest, **Gulmarg** WLS, **Hirpora** WLS, and **Overa-Aru** WLS. Intesar Suhail (*pers. comm.*) has noted confirmed breeding records of the bird from Dachigam National Park (19.06.1998; 05.05.2001) and more recently (15.06.2012) from Hajin Conservation Reserve, Tral in south Kashmir.

Ecology: The Kashmir Flycatcher breeds in summer in the Northwest Himalaya, in Kashmir and the Pir Panjal Range between 1,800 and 2,400 msl, and even up to 2,700 msl (Bates & Lowther 1952, Henry 1955, Roberts 1992). It is reported to be common in **Overa** WLS in Kashmir (Price & Jamdar 1990), mainly found in temperate, mixed deciduous forests, particularly those comprising Hazel *Corylus*, Walnut *Juglans*, Cherry *Prunus*, Willow *Salix*, and Parrotia *Parrottetia* sp., with a dense, shrubby understorey. In **Hajin** Conservation Reserve, **Tral** (South Kashmir), Intesar Suhail (*pers. comm.*) observed the bird in a mixed broadleaf forest of Maple *Acer* sp., Horse Chestnut *Aesculus indica*, and *Ulmus* sp. with undergrowth of *Viburnum* sp.

The Kashmir Flycatcher is insectivorous (Henry 1955). Zarri & Rahmani (2004) reported that it feeds mainly on insects, including small butterflies, moths, grubs, larvae, caterpillars, and also on earthworms. The birds were often seen coming down to buffalo dung and digging out grubs from the heap. Size of the food item varied from a few millimetres to nearly 12 cm (earthworm). The bird usually fed very close to the ground, about 1–2 m. For details, see Zarri & Rahmani (2004).

Baker (1924) writes, "The nest is cup-shaped and is made of moss and dead leaves, mixed, more or less, with scraps of grass, chips of leaves and dead wood, hairs and feathers...". Further, he writes "seems to be invariably placed in holes in trees...". Bates & Lowther (1952) found nests in holes and small slits, often leading vertically downward, and woodpecker holes were also appropriated. According to Roberts (1992), if its breeding habits resemble the Red-breasted Flycatcher, then most of the nest building and incubation must be done by the female alone. Most individuals leave the breeding grounds in September, arriving in Sri Lanka in October and departing again in late March.

Zarri & Rahmani (2004) have shown that in its wintering areas the Kashmir Flycatcher selects open areas with dense grass cover and low shrubs, and avoids areas with high tree density. Wattle plantations with openings created by transmission lines provide such a habitat, but areas like this are generally subjected to much grazing and lopping pressure. The species avoids thick sholas and may have adjusted to the wattle plantation openings and edges. Pair-bonds appear to be maintained throughout the winter, and winter territories are occupied in successive years, suggesting strong winter site fidelity (Zarri & Rahmani 2004). According to Trevor Price (*in litt.*) this is perhaps the first time a migratory passerine has been reported in pairs in the winter, and this needs a more detailed study with marked birds. Kylin (2002) saw a pair on August 3, 2000, at 1,700 msl in a tea plantation at the Dambatenne Estate (6° 46' N, 80° 59' E) in the southern part of the Sri Lankan highlands "displaying distinct nesting behaviour...". From his 20 minutes observation of the pair, Kylin deduces that they were probably nesting, although no attempt was made by him to locate the nest. However, Zarri & Rahmani (2004) found that the Kashmir Flycatcher lives in pairs even in its wintering quarters, and territories are occupied in successive years. Therefore, sighting a pair does not necessarily mean that there is nesting.

The pairs observed by Zarri & Rahmani (2004) were very parochial and seen in their territories throughout the winter. Call is a characteristic *whip, whip, whip*... and *chrit..rr..chrit*, sometimes heard at a considerable distance.

Threats: In India, there is no threat from poaching or killing for this tiny bird, nor is trade a major threat (Rajat Bhargava *pers. comm.* 2009) but habitat changes and anthropogenic pressure on its wintering quarters in the Nilgiris are serious threats that may have already affected the species' existence. Commercial and non-commercial timber extraction, conversion of land to agriculture, fuel wood collection, and livestock grazing are common in all its wintering areas in India. In the Nilgiri Plateau, the poor economic feasibility of existing wattle plantations has led to increased rates of clearance, bringing about reduction in the areas suitable as wintering habitats (Zarri & Rahmani 2004).

Conservation measures underway: Although it is not listed by name in the Indian Wildlife (Protection) Act, 1972, shooting and trapping of all wild bird species is prohibited under the Act. All members of Muscicapidae are listed in Schedule IV of the Act. It is also listed under CMS Appendix II. In India, it is found in and around some protected forests and IBAs. In Kashmir, it breeds in Overa WLS and Dachigam NP, where the habitat is well protected. It also occurs in a few protected areas in Sri Lanka.

RECOMMENDATIONS

(1) Conduct surveys across its breeding and wintering ranges to establish its current population status.
(2) Study its breeding behaviour, habitat requirements, and ecology in Overa WLS, and wintering ecology near Avalanche and Mukurthi NP.
(3) Conduct ringing and colour banding studies to trace its short distance and long distance movement, and also recruitment, age structure, and longevity.
(4) Identify key breeding and wintering sites and campaign for their protection where necessary.
(5) Provide increased support and resources for more effective protected area management within its breeding range.

Ferruginous Duck
Aythya nyroca (Güldenstädt 1770)

DHRITIMAN MUKHERJEE

BirdLife International (2013) considers the Ferruginous Duck Near Threatened due to rapid decline of its numbers in some parts of its wide range.

Field Characters: An overall dark chestnut diving duck, *c.* 41 cm, with a large oval white patch on the belly (clearly visible in flight), with conspicuous white eyes in male visible clearly at short distance. Female is like the male but duller, with brown eyes. Both sexes are slightly darker on the back. Juvenile similar to adult but belly and undertail are grey-buff. In flight, a broad white wing-bar extends to outer primaries. Locally it is known as *harawöt*.

Distribution: The Ferruginous Duck or White-eyed Pochard is widely distributed in the Palaearctic region from western Europe to western Mongolia. There is an isolated breeding population in Libya. In north India, it is a common winter migrant (Ali & Ripley 1987, Grimmett *et al.* 1998). Earlier it was found breeding in some wetlands of Kashmir such as Hokarsar, Anchar, and Haigam (Bates & Lowther 1952), but we do not have recent confirmed records of breeding. Earlier it used to breed in such vast numbers that collection of the eggs of this duck and of the Mallard, and bringing them into Srinagar by boat for sale, was a regular and profitable profession for a number of people living in the vicinity of their breeding haunts (Baker 1921).

Bates & Lowther (1952) write "More numerous than the Mallard, being distributed in some numbers throughout those jheels of the Vale which have plenty of reedy cover and water-plants. Can be considered fairly common on the State *rukhs* such as, for instance, Hokra, the northern end of the Anchar Lake, and Haigam. Their numbers are considerably increased in the cold weather by fresh

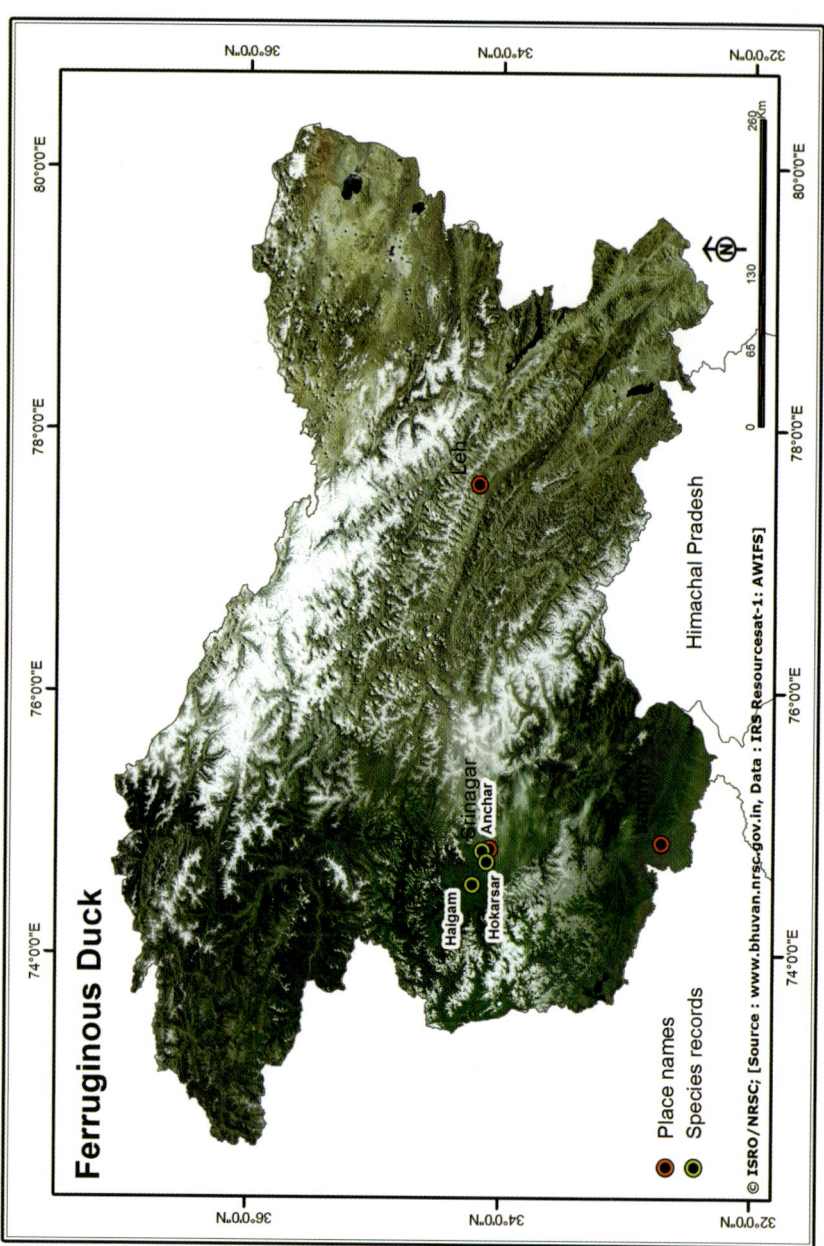

Ferruginous Duck

Leh

Srinagar
Anchar
Halgam
Hokarsar

Himachal Pradesh

© ISRO/NRSC; [Source : www.bhuvan.nrsc.gov.in, Data : IRS-Resourcesat-1: AWIFS]

Place names
Species records

arrivals from the north-west." Unfortunately, it is now uncommon in the Valley and there is no known breeding record in recent years that we know of.

Ecology: In India, it can be seen in shallow ponds, pools, and marshes near vegetated shoreline, large marshes, wetlands, and sometimes in the rivers. Prefers shallower and more vegetated areas than other *Aythya* species and seldom sits out in open water. It feeds on seeds, roots, and the green parts of aquatic plants and also on insects, worms, molluscs, crustaceans, amphibians, and small fish. It often feeds at night, and will upend (dabble) for food as well as the more characteristic diving. According to del Hoyo *et al.* (1992), it starts breeding by April and May in Central Europe, singly or in loose groups, making its nest with leaves, stems, grass on ground with thick vegetation or in reed beds. It lays 8–10 eggs, incubating them for 25–27 days, and the chicks become sexually mature in one year. It is a gregarious species, forming large flocks in winter, often mixed with other diving ducks.

According to Bates & Lowther (1952), in the Kashmir Valley, it nests rather later than the Mallard, throughout May, though eggs may be found up to the end of June. They write "The nest is a circle of coarse grass and rush leaves with a somewhat finer lining, and as incubation proceeds the bird adds considerable quantities of its own dark-brown down."

Threats: The Ferruginous Duck is threatened by the degradation and destruction of well-vegetated shallow pools and other wetland habitats (e.g., changes in the vegetation community, disruption of water regimes, siltation, and increased water turbidity) as a result of excessive drainage and water abstraction, peat extraction, eutrophication (from inadequate sewage treatment and nutrient run-off), oil pollution, dam and barrage construction, building of infrastructure on floodplains and river canalisation (BirdLife International 2001, 2013). Various threats in different countries, or parts of its wintering and breeding ranges, are also listed in the same sources. In India, it is mainly threatened by habitat destruction and modification, and by trapping and poisoning.

Conservation measures underway: In India, it is protected under the Indian Wildlife (Protection) Act, 1972, which bans its hunting, trapping, trading, and poaching. It is listed in Schedule IV of the Act.

RECOMMENDATIONS

Rahmani (2012) has given India-specific recommendations. Here we give suggestions for its protection in Jammu & Kashmir:

(1) Survey to find out its status and distribution in the state, particularly its breeding areas.
(2) Development of State Wetland Conservation Act for protection of wetlands.
(3) Better protection to non-protected wetland IBAs.
(4) Regular monitoring of all waterfowl, particularly threatened species.

Black-necked Stork
Ephippiorhynchus asiaticus (Latham 1790)

DHRITIMAN MUKHERJEE

BirdLife International (2013) justifies listing the Black-necked Stork as Near Threatened, as this species has undergone a moderately rapid overall population decline, which is projected to continue, and has a moderately small population. The global population estimates vary between 10,000 and 20,000.

Field Characters: A large bird, standing 130 to 150 cm tall, with bright red legs, white body, extensive black on the wings and tail, and a conspicuously glossy, iridescent black head and neck, with a large black bill. The young look like a washed-out version of the parent, with dull brown replacing glossy black parts, and dirty white replacing pure white. Genus *Ephippiorhynchus* is unique among storks in exhibiting sexual dimorphism: iris dark brown in male and yellow in female. Like most storks, the Black-necked Stork flies with its neck outstretched, not retracted like a heron. The wingspan is up to 230 cm.

Distribution: The Black-necked Stork is found across the Indian plains, widespread but not common. It is also seen regularly on coastal wetlands and mangrove swamps, and even breeds on tall mangrove trees in Gujarat. Extralimitally, it is found in Pakistan, Nepal, Bangladesh, and Sri Lanka.

Black-necked Stork

Place names ●
Species records ●

© ISRO/NRSC; [Source : www.bhuvan.nrsc-gov.in, Data : IRS-Resourcesat-1: AWIFS]

Leh

Srinagar

Gharana

Himachal Pradesh

PUSHPINDER SINGH JAMWAL

In J&K, Jammu is the only region where Black-necked Stork is reported, particularly from the famous Gharana Wetland

Distribution records from India were earlier compiled by Rahmani (1989) and recently updated by Rahmani (2012). Here we give records specific to Jammu & Kashmir. In the state, the species is regularly seen only in the **Gharana** Wetland Reserve near the Indo-Pak border where a few birds have been recorded every year by the Wildlife Department field staff and other visitors. Ashfaq Ahmed Zarri and Pankaj Chandan recorded two to three birds during their different visits in the winter to Gharana from 2006 to 2011. During an ongoing study of the birds of Gharana Wetland, researchers of WWF-India sighted 6–8 individuals on several occasions during different visits in winter. They also recorded 14 Black-necked and *c.* 50 Woolly-necked Stork around the agricultural fields near **Gharana** Wetland Reserve on December 5, 2012 (Pushpinder Singh Jamwal & Rohit Rattan *pers. comm.* 2012).

Ecology: The Black-necked Stork inhabits large marshes and jheels, and margins of large rivers and brackish lagoons, where it feeds on fish, frogs, snakes, small turtles, injured and unwary birds, and any animal which it can swallow (Ishtiaq *et al.* 2010). The nest is built on large trees, mostly near water. If undisturbed, the same tree is used in following years. Ishtiaq *et al.* (2004) recorded nest fidelity in Keoladeo NP. Pairs spend considerable time on the nest, where mating also occurs. The female lays two to four eggs. Both parents incubate and raise the chicks. Generally one to three chicks are raised (Sundar 2003), but in such habitats as Dudhwa NP and Kishanpur WLS in Uttar Pradesh, where food is plentiful, it is not uncommon to see four juveniles with parents in some years. Ishtiaq (1998), during her studies in Keoladeo NP, found three juveniles raised by a pair in 1996–1997. Similarly, Bhatt (2006) reported a pair with four fledged juveniles in Marine NP, Gujarat, and Sundar *et al.* (2007) found four chicks in Etawah and Mainpuri with some pairs. Detailed studies have been done by Sundar (2003), Ishtiaq (1998) and Maheswaran (1998).

Threatened Birds of Jammu & Kashmir

Threats: The main threat to the Black-necked Stork is destruction and degradation of its habitat, and overfishing. Hunting, at least in J&K, is not the main problem, but trapping for zoos was at one time the major problem in India (Rahmani 1989).

Conservation measures underway: This stork is listed in Schedule I of the Indian Wildlife (Protection) Act, 1972 and also included in CITES Appendix I. In India it occurs in a number of PAs/IBAs where detailed research has been done. Despite the fact that the species has very limited distribution in the state of Jammu & Kashmir, and that too in unprotected areas, there has been no effort to conserve it, compared to the efforts made for other ecologically similar wetland birds in J&K.

RECOMMENDATIONS

BirdLife International (2013) has given general recommendations for all range countries of this species. For J&K and the other range states, we make the following recommendations:

(1) Since accurate information with regard to their number and distribution is lacking in Jammu & Kashmir, efforts should be made to document their present status and distribution.

(2) More attention should be given to the Black-necked Stork populations outside PAs and IBAs.

(3) Strict control should be implemented on the use of harmful pesticides, particularly near wetlands.

(4) A nation-wide programme should be implemented to ring/band (both colour and aluminium) juveniles of Black-necked Stork every year to track their dispersal movements, which will give a better estimate of their population trends.

(5) The degree to which habitat fragmentation or loss might affect Black-necked Stork is related, in part, to their movements, about which almost nothing is known. Tracking individuals using satellite telemetry would greatly assist in conservation efforts.

(6) Initiation of an active advocacy programme aimed at farmers on the importance of wetland birds and their protection.

(7) Total implementation of the ban on hunting and trapping of these birds.

(8) Involvement of IBCN members in monitoring local Black-necked Stork nesting and foraging sites. Since Gharana is facing severe management problems due to socio-economic reasons and its precarious location amid agricultural fields on one side and the Indo-Pak border on the other, the Department of Wildlife (Protection) should consider the special ecological requirements of this species while working out a scientific management plan.

(9) Make Black-necked Stork, along with Sarus Crane, an icon of healthy wetlands.

Black-headed Ibis
Threskiornis melanocephalus (Latham 1790)

DHRITIMAN MUKHERJEE

According to BirdLife International (2013), the Black-headed Ibis, in common with most large wetland species in Asia, is undergoing a population reduction. It faces the entire gamut of threats, from hunting and disturbance at breeding colonies to drainage and conversion of foraging habitats, to agriculture. It consequently qualifies as Near Threatened.

Field Characters: A large, domestic hen-sized white bird, with black neck, naked black head, and long stout downcurved black bill. Legs and feet are also black. Adult breeding birds have white lower neck plumes, a variable yellow wash on the mantle and breast, and grey on scapulars and elongated tertials. Immature birds have a white chin and neck, naked face and bare skin around the eyes, while the rest of the head and neck are feathered. Sexes alike.

Distribution: The Black-headed Ibis, also called White Ibis, is widespread and locally common all over the wetter parts of western India, becoming scarce in the east. Even in the driest parts (e.g., Thar Desert, Kutch) it is now spreading due to the development of canal irrigation. It is resident, nomadic, and locally migratory, depending upon the availability of water.

Black-headed Ibis

Leh

Srinagar

Gharana

Himachal Pradesh

260 Km

130

65

0

● Place names

● Species records

© ISRO/NRSC; [Source : www.bhuvan.nrsc.gov.in, Data : IRS Resourcesat-1: AWIFS]

During a bird survey of the Jammu region by a team from WWF-India, four of these birds were recorded and photographed on September 26, 2012 at **Gharana** Wetland Reserve near the Indo-Pak border in Kathua district (Pushpinder Singh Jamwal & Rohit Rattan *pers. comm.* 2012). These ibises were not seen thereafter during regular fortnightly surveys.

Ecology: The Black-headed Ibis inhabits various wet habitats, from paddy fields, freshwater marshes, lakes, rivers, and flooded grasslands, to tidal creeks, mudflats, salt marshes, and coastal lagoons, from lowlands to 950 msl. It is gregarious, found with waders like storks, egrets, spoonbill, and other small waders. It is always found close to water. Feeds almost entirely on animal matter, fish, frogs, aquatic insects, crustaceans, and worms, the last two generally probed out of the slush by its downcurved bill.

It nests during the monsoon, colonially with other species in heronries, in partially submerged thorny trees to avoid ground predators. A platform nest is built, and two to four eggs are laid. Incubation and chick rearing are shared by the male and female. The House Crow *Corvus splendens* is a major predator of eggs and small chicks, while nestlings are attacked by eagles. Where not molested, it nests on trees growing even in crowded villages, sometimes away from water, with other colonial nesters such as Painted Stork, Grey Pelican, and various egrets.

Threats: It suffers from the same threats as all species dependent on natural wetlands are suffering in South and Southeast Asia: drainage, disturbance, pollution, agricultural conversion, hunting, and collection of eggs and nestlings. All these factors have led to the decline of its populations in some countries.

Conservation measures underway: It is listed in Schedule IV of the Indian Wildlife (Protection) Act, 1972 so its hunting in India is totally prohibited. It is found in many PAs/IBAs, and also gets protection due to religious and traditional practices in many areas (e.g., wetlands associated with temples).

RECOMMENDATIONS

(1) Strictly enforce ban on trapping and prevent poaching, particularly during the species' breeding season.

(2) Conduct surveys in the various wetlands of Jammu region and their associated areas, in order to obtain data on their current status.

(3) Conduct surveys involving members of BNHS, IBCN, other conservation organizations, and civil society, to identify and protect important breeding colonies. Subsequently, conduct surveys of identified breeding colonies every two years to monitor the population trend.

(4) Study the impact of pesticides on the food chain of this species.

(5) Conduct ringing, colour banding, and satellite tracking studies to determine its movements.

Black-tailed Godwit
Limosa limosa (Linnaeus 1758)

BHASMANG MEHTA

Although the Black-tailed Godwit is widespread and has a large global population, its numbers have declined rapidly in parts of its range owing to changes in agricultural practices. Since 2006, it has been in the Near Threatened list of BirdLife International. It was thus categorised due to an estimated decline of around 25% of its population during the preceding 15 years (BirdLife International 2013).

Field Characters: One of the large waders of India, it measures 40–44 cm, with a distinctive long bill on a relatively small head, long neck, and long legs. In winter, the colour of its forebody is pale grey-brown, turning dull pink-chestnut by March-end or April when the birds migrate to their breeding areas. In flight, a striking white wing-bar and rump, and black tail are diagnostic. The female is *c*. 5% larger than the male.

Distribution: The Black-tailed Godwit is a fairly common winter migrant in India, more so in freshwater marshes and jheels than on the coast. It spreads over the entire Indian subcontinent. Rahmani (2012) has compiled all major records from India, whereas here we give records from Jammu & Kashmir.

Small flocks of this species have been recorded at **Gharana** Wetland Reserve towards the end of their wintering season in the last few years (Ashfaq Ahmed Zarri & Pankaj Chandan *pers. comm.*). The bird had not been recorded during the fortnightly surveys being conducted by WWF-India's research team at Gharana during winter, uptil December 25, 2012. Presumably, Gharana is a stopover site for some small flocks on their way back to their summer breeding grounds. Pfister (2004) described the species as a rare passage migrant during autumn (August to September) through high altitude wetlands (4,650 msl) in eastern Ladakh. Single birds were observed at the **Tso Kar** marshes (Pfister 2004).

Black-tailed Godwit

Leh

Tso Kar

Srinagar

Gharana

Himachal Pradesh

Place names
Species records

© ISRO/NRSC; [Source : www.bhuvan.nrsc.gov.in, Data : IRS-Resourcesat-1: AWIFS]

0   65   130   260
Km

Ecology: The Black-tailed Godwit is a winter migrant in India, with the first birds arriving by the last week of August or early September, and becoming well dispersed all over India by late November and December. Very gregarious, sometimes found in flocks of tens of thousands (e.g., in Chilika Lake in Orissa) foraging on soft mud and ooze for small invertebrates. Sometimes it is seen solitarily or in small parties of 5–10 individuals on roadside puddles or village ponds. It feeds on tiny molluscs, crustaceans, worms, and seeds of grass and marsh plants.

It is very silent in winter, except for a low trisyllabic *wit-wit-wit* or *quick-quick-quick* uttered when taking off from the ground (Ali & Ripley 1987). During the breeding season, it makes high-pitched, nasal, rather strident calls, the most common of which is *weeka-weekaweeka* (BirdLife International 2013). According to BirdLife International (2013), it breeds from April to mid-June in loose, semi-colonial groups of up to three pairs per hectare. It breeds in low-lying wet grasslands, grassy marshland, raised bogs and moorland, lake margins, and damp grassy depressions on steppes. Secondary habitats such as wet meadow, pasture, damp areas around fish-ponds and sewage ponds, and salt-water lagoons are also used.

Threats: In breeding areas, loss of nesting habitat owing to wetland drainage and agricultural intensification, hunting (France), and poaching are the major threats (BirdLife International 2013). In its winter quarters, the same factors operate to a different degree. Drainage of wetlands, afforestation of shallow wetlands and mudflats, poaching and pollution are the major threats. Climate change is an impending threat that would play a major role in further decline of this species. For details of threats, see BirdLife International (2013). It is one of the target waders of poachers, as it is good to eat. Therefore it is caught in large numbers by trappers in India (Daniel *et al*. 1999).

Conservation measures underway: Like all waders, it is protected under the Indian Wildlife (Protection) Act, 1972, and listed in Schedule IV of the Act. It occurs in a large number of IBAs/PAs. An EU Management Plan for 2007–2009 has been adopted. It is also among the species to which the Agreement on the Conservation of African-Eurasian Migratory Waterbirds (AEWA) applies. Intensive management of breeding habitat has been carried out in some west European countries, and a number of agri-environment schemes focus on this species, though results have been mixed (BirdLife International 2013).

RECOMMENDATIONS

(1) Strict ban should be implemented on trapping and shooting of all waders in and around Gharana Wetland Reserve and other wetlands.

(2) Develop policies to protect wetlands, estuaries, lagoons, and mudflats by promulgating a Wetland (Conservation) Act.

(3) Develop State Bird Monitoring Scheme involving forest officials, NGOs, university departments, and members of civil society.

(4) Prevent afforestation of wetlands, mudflats, and swamps, which are the wintering grounds of the Black-tailed Godwit and numerous other waders.

Eurasian Curlew
Numenius arquata (Linnaeus 1758)

BirdLife International (2013) states that Eurasian Curlew remains common in many parts of its range, and determining population trends is problematic. Nevertheless, decline has been recorded in several key populations and overall a moderately rapid global decline is estimated. As a result, the species has been uplisted to Near Threatened.

Field Characters: A large (55–58 cm), pale, sandy-brown wader with a long downcurved bill. The mottled or scalloped brown plumage with whiter belly and undertail, and fine, short, black streaks on underside are typical. In flight, it shows pointed whitish rump and barred tail as well as mottled whitish underwings, while the outer primaries are dark in contrast. The flight is slow and gull-like. Sexes are alike. Juveniles are somewhat darker, with finer black streaks on breast and very few on the abdomen.

Distribution: This winter visitor to India is mainly seen in coastal areas, but also frequents large jheels and rivers. It arrives by end August or early September, and generally leaves by end April, but is sometimes seen up to May and even June. In south India, mainly on the Rameswaram coast and in the Gulf of Kutch, some birds over-summer (S. Balachandran *pers. comm.* 2013).

In the Kashmir Valley, it has been recorded from the **Wular Lake** by Walter Lawrence, and also from the **Veshau** river in south Kashmir where he shot "one out of a party of three" in 1893 (Lawrence 1895).

In Ladakh, it is an occasional but regular passage migrant during autumn (late July to early September) through the depths of the valleys of western and central Ladakh, and the eastern high altitude plains (up to 4,650 msl) (Pfister 2004). Typical areas of sightings reported by Pfister (2004) are **Kargil** (August), the **Shey**

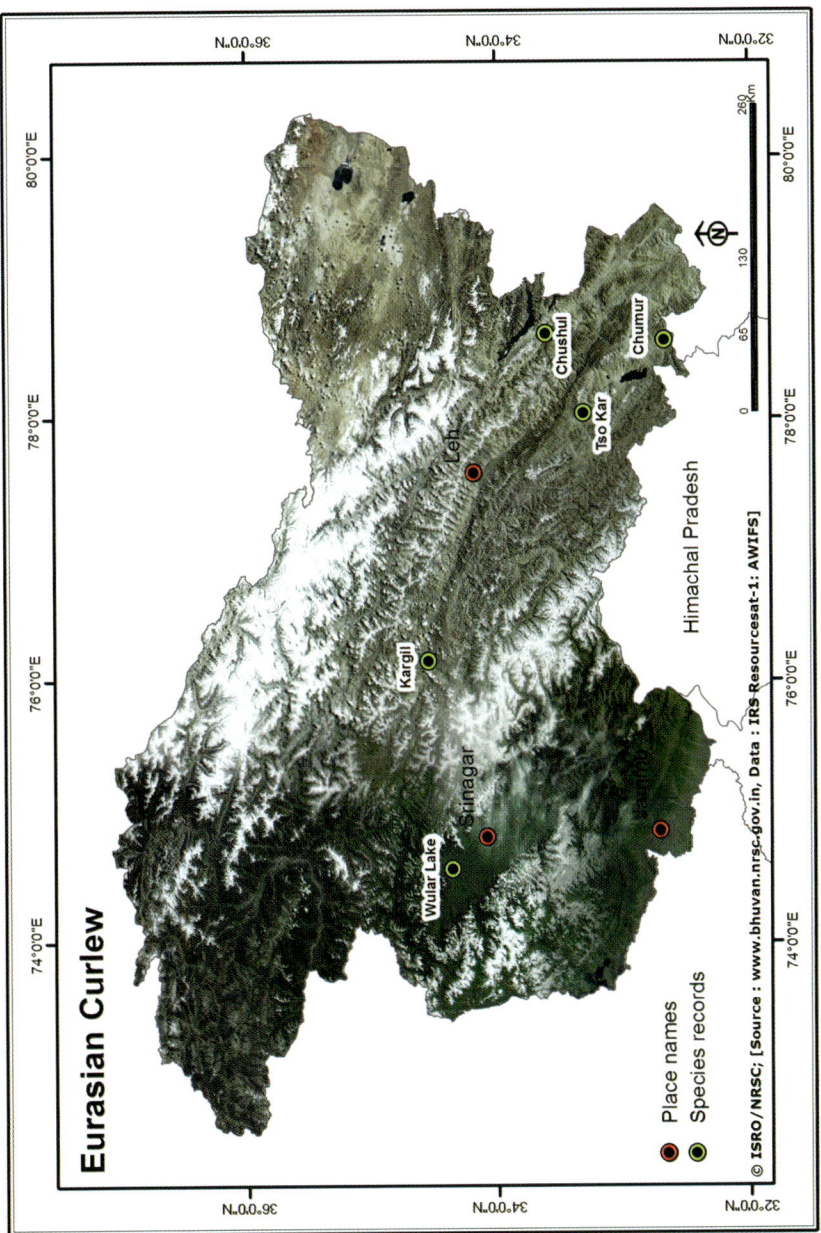

Eurasian Curlew

Chushul
Chumur
Tso Kar
Leh
Kargil
Srinagar
Wular Lake
Himachal Pradesh

● Place names
● Species records

© ISRO/NRSC; [Source : www.bhuvan.nrsc.gov.in, Data : IRS Resourcesat-1: AWIFS]

Tikse marshes (September), and **Lam Tso/Chumur** (August), but it is seen more regularly in **Startsapuk Tso** and **Tso Kar** (July–September) and **Hanle** (July).

Pankaj Chandan and Nisa Khatoon reported sighting this species from **Hanle** and **Chushul** marshes in Leh during their surveys in 2013. Naoroji & Sangha (2003), during their surveys in Ladakh region between 1997–2003, reported sighting it in Ladakh as a passage migrant.

Ecology: The Eurasian Curlew is found singly or in small scattered flocks on coastal swamps, intertidal zones, beaches, edges of large rivers, jheels, and reservoirs, along with assorted waders and egrets. It feeds on annelid worms, aquatic insects and their larvae, crustaceans, molluscs, spiders, as well as berries and seeds. It can be seen feeding on damp grasslands, sometimes inland. It is reported to feed on small fish, amphibians, young lizards, young birds, and small rodents (del Hoyo *et al*. 1996). Males are more often reported feeding in inland grasslands than females. It does not breed in India, but in the temperate region it has been found breeding on upland moors, peat bogs, swampy or dry heaths, fens, open grassy or boggy areas in forests, damp grasslands, meadows, dune valleys, and coastal marshlands (del Hoyo *et al*. 1996).

Threats: BirdLife International (2013) has described the main threats in its breeding areas. These include loss of breeding and foraging habitats, as well as hunting and trapping in wintering areas. Another major threat is the conversion of mudflats to mangrove plantations, which is happening in Gujarat, Maharashtra, Kerala, and other states. Extensive trapping of all waders including Eurasian Curlew occurs in Bihar, Uttar Pradesh, Tamil Nadu, Andhra Pradesh, and parts of Gujarat (Ahmed 2002). The birds are sold directly to roadside hotels (*dhaba*s) and are brazenly advertised on the menu!

Conservation measures underway: Like all resident and migrant waders, it is protected under the Indian Wildlife (Protection) Act, 1972. It is listed in Schedule IV of the Act. It is also listed in Annexure II/2 of the EU Birds Directive. The European Commission instituted a management plan for the species which was updated for 2007–2009 (BirdLife International 2013). This curlew occurs in many PAs and IBAs in India, and throughout its range features in several national monitoring schemes. This is one of the species to which the Agreement on the Conservation of African-Eurasian Migratory Waterbirds (AEWA) applies.

RECOMMENDATIONS

BirdLife International (2013) suggests measures for a realistic management plan for the Eurasian Curlew with key conservation targets. For J&K, we suggest the following measures:

(1) Strict control on trapping and shooting.
(2) Regular monitoring of Eurasian Curlew populations through IBCN and AWC network.
(3) Research and survey initiatives within J&K in its potential habitats to establish its current status.

Cinereous Vulture
Aegypius monachus (Linnaeus 1766)

ASAD R. RAHMANI

The Cinereous Vulture has a moderately small population which appears to be suffering a decline in its Asiatic strongholds, despite the fact that in parts of Europe its numbers are now increasing. Consequently it has been listed as Near Threatened (BirdLife International 2013).

Field Characters: A huge vulture, between 98 and 107 cm, mostly black or dark brown, with broad wings and a short, often slightly wedge-shaped tail. The adults are dark brown, while juveniles are blackish, with dark crown, ruff, and upper breast, contrasting with paler adults. It has an almost naked head, with a massive bill, with crown, lores, and cheeks covered with black fur-like feathers and down. Bare skin of head and neck bluish grey. Sexes alike, but female is larger (2–4%) and heavier (*c.* 7%) than male. It is one of the largest vultures found in India. In its wide distribution range, size increases from west to east, with the Mongolian and Chinese birds larger than European birds (Ferguson-Lees & Christie 2001).

It has a typically unfeathered bald vulture head (which is actually covered in fine down) and dark markings around the eye, giving it a menacing skull-like appearance. The beak is brown, with a blue-grey cere, while legs and feet are grey.

Distribution: The Cinereous Vulture has a wide range from southern Europe, north Morocco, Algeria, Sudan, the Middle East, Central Asia, to Mongolia and

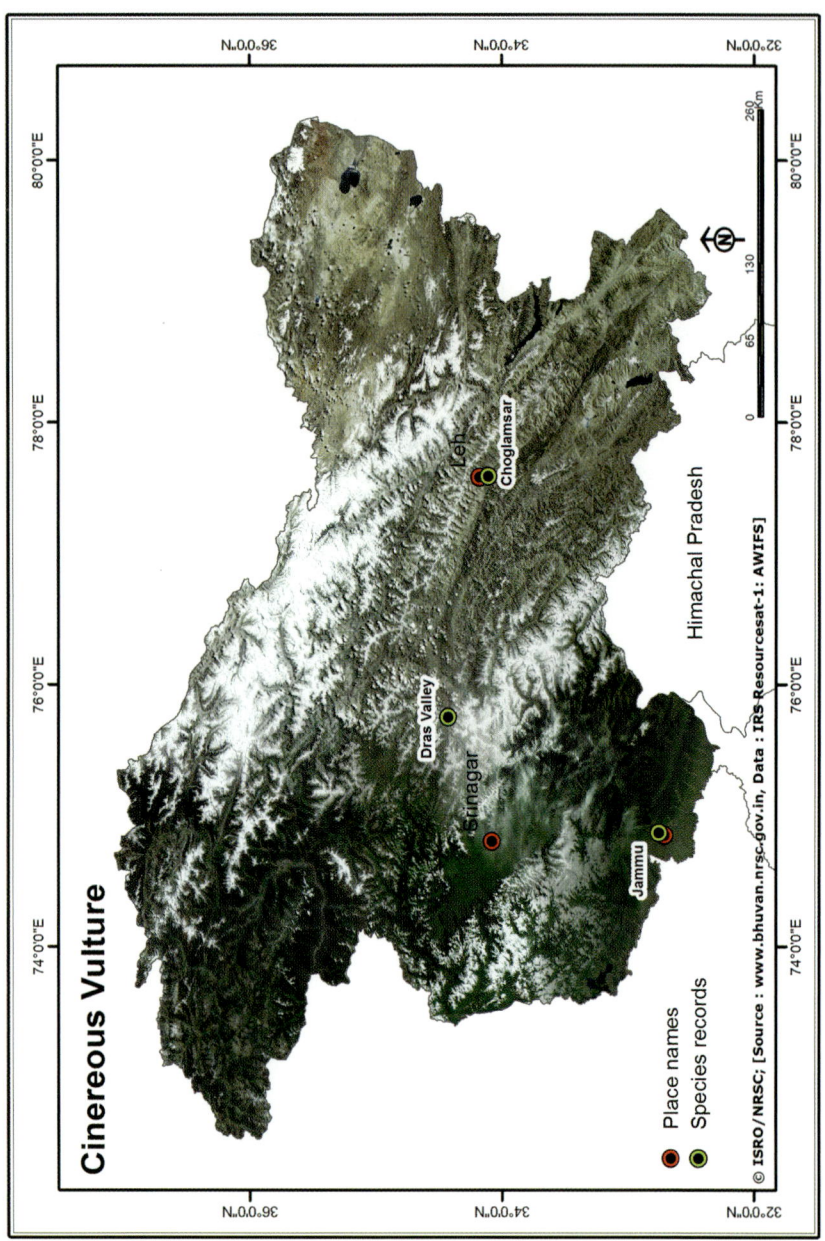

Cinereous Vulture

Himachal Pradesh

Leh
Choglamsar

Dras Valley

Srinagar

Jammu

● Place names
● Species records

© ISRO/NRSC; [Source : www.bhuvan.nrsc.gov.in, Data : IRS Resourcesat-1: AWIFS]

east China. In India it is seen only in winter, mainly in the plains of Gujarat (Kutch, Saurashtra), Rajasthan (more frequently in the Thar Desert), Punjab, Haryana, western Uttar Pradesh, Uttarakhand, Maharashtra (with records from Dhulia), Assam, and Jammu & Kashmir. It is quite common in Uttarakhand. Pfister (2004) reported sightings from Choglamsar near **Leh**, and **Dras Valley** in Kargil. Williams & Delany (1996) reported it as a passage migrant in Ladakh. Naoroji & Sangha (2003) have not listed the species in the checklist made from their extensive raptor surveys during 1997–2003. An injured bird captured from **Kathua** district in Jammu by the Department of Wildlife (Protection) field staff is under rehabilitation in Manda Zoo in Jammu for the past several years.

Ecology: In India, it is seen singly or in twos or threes, roosting early in the morning on large trees or seated on sand dunes or mounds, commanding a view of the surroundings. In its breeding areas, it is also found in forested hills, as also alpine grasslands and steppes. Generally a solitary nester, it sometimes nests in very loosely associated colonies. It feeds mainly on large mammal carcasses, but is also reported to feed on stranded turtles and large dead birds, e.g., Painted Stork. It dominates the jostling rabble of other vultures at a carcass, and can be quite aggressive. It is equipped with a powerful bill that can tear open tough carcass skins. It is generally very silent but while feeding, it croaks, grunts, and hisses.

Threats: BirdLife International (2013) has listed two major threats to Cinereous Vulture: direct mortality caused by humans (either accidentally or deliberately) and decreasing availability of food. Due to its large size and long life, it is also caught for zoos (Roberts 1991), while in China it is trapped or shot for feathers.

In India, hunting is not a likely threat as this bird is left alone, and lack of food is also not a problem, particularly where the bulk of winter birds are found in the arid Thar Desert. However, diclofenac poisoning through livestock carcasses could be a major threat, though it is still not confirmed. In Tibet, where it has been traditionally protected by Tibetans, increasing use of rodenticide by Chinese colonisers has a serious impact on this species and other carrion-eaters.

Conservation measures underway: It is protected under the Indian Wildlife (Protection) Act, 1972, and listed in Schedule IV of the Act.

RECOMMENDATIONS

(1) In India, where mostly juveniles birds are recorded wintering, they are known to wander long distances. Thus, every winter a number of starving and dehydrated individuals are rescued. The birds recover within a few days when food and water is provided. There is a need to create awareness about the species, and people should be encouraged to send sick and injured birds to rescue centres. A large number of juveniles could be saved by providing water and care for a few days.

(2) Thorough post-mortem examination should be carried out on any dead bird found, the tissues should be checked for the presence of diclofenac, and the results reported.

(3) Regular surveys should be carried out throughout its known wintering range to determine its wintering population trends.

European Roller
Coracias garrulus Linnaeus 1758

SACHIN RAI

According to BirdLife International (2013), the European Roller apparently underwent a moderately rapid decline across its global range and was consequently uplisted from Least Concern to Near Threatened in 2005. Its decline has been most pronounced in northern populations, and if similar trends are observed elsewhere in the species' range, it may warrant uplisting to Vulnerable. Its present numbers could be between 100,000 and 500,000, but it is decreasing.

Other names: Eurasian/Blue/Common Roller, or The Roller.

Field Characters: A pale roller with turquoise head and underparts, and tawny mantle. Wings blue with black flight feathers, clearly visible in flight. Large head and black bill, upper mandible hooked at the tip. Sexes similar, while the juvenile is a drab version of the adult.

Distribution: The European Roller has an extensive breeding range in Europe, northern Africa (Morocco), the Middle East (Iraq), Iran, Central Asia, and northwest China. BirdLife International (2013) has shown that it is undergoing population decline in many areas and its overall European decline exceeded 30% in three generations (15 years). However, there is no evidence of any decline in Central Asia. Should these populations exhibit further decline, the species may require uplisting to Vulnerable. The European Roller is a long-distance migrant, wintering in southern Africa in two distinct ranges, from Senegal east to Cameroon, and from Ethiopia westwards to Congo and south to South Africa.

In India, the European Roller breeds only in the Kashmir Valley, elsewhere it is a passage migrant, through northwest India (Punjab, Haryana, Rajasthan,

European Roller

Place names
Species records
© ISRO/NRSC; [Source : www.bhuvan.nrsc.gov.in, Data : IRS-Resourcesat-1: AWIFS]

Leh
Ladakh
Srinagar
Anantnag
Jammu
Himachal Pradesh

Like its cousin, the Indian Roller *Coracias benghalensis,* the European Roller is also found in lightly wooded country and nests in tree holes

Gujarat, western Madhya Pradesh, and western Maharashtra) from end August to September. Some individuals have been reported lingering till October.

Pfister (2004) described the species as a rare passage migrant in **Ladakh**, during autumn (September). It may be met with along the Indus, but is also seen during summer in western Ladakh as a vagrant from Kashmir. In the Jammu region, we do not have a confirmed sighting.

Lawrence (1895) found the species "fairly common" in Kashmir "from April to the end of September". He "chiefly noticed it near Islamabad (now **Anantnag**), Kulgam and in the lower parts of the Sind and Liddar and in the Lolab and Kamraj". Intesar Suhail (*unpubl.* 2012) recorded the nesting of this species from two areas in **Anantnag** district. A pair was observed nesting in a hole in a concrete retaining wall at the newly constructed Anantnag railway station. The nest was about 3 m above ground level, beside a busy car park. Another pair was recorded nesting in a soil bank near village Panzpora, Sangam.

Ecology: In the Kashmir Valley, it breeds in open wooded vales, parks, and cultivated areas, nesting in a natural tree hollow from 3 to 10 m, but "perhaps the most favoured site is a cavity in a sand-bank" (Bates & Lowther 1952). The clutch is four to five eggs, and chicks hatch by end June or early July (in Kashmir). Both parents incubate and rear the chicks. They forage in cultivated areas, fallow fields, and open woodlands for large insects, ground beetles, lizards, frogs, and small injured birds that are easily caught.

Threats: In India, the main threat to this species (and most insectivorous birds) is through direct pesticide poisoning, or indirectly through decrease in the insect supply. It is also killed during migration in some Mediterranean countries and hundreds, perhaps thousands, are shot for food in Oman every spring (del Hoyo *et al*. 2001). In Europe, it is known to be sensitive to loss of hedgerows and riparian forest which provide essential habitats for perching and nesting (BirdLife International 2013).

As it happens, the Indian Roller *Coracias benghalensis* is considered an incarnation of Lord Shiva, and during the Hindu festival of Dassera sighting this bird is considered auspicious. In early October, the Indian Roller is caught in large numbers for the release trade on Dassera, where people pay to release their "sacred bird". The hapless European Roller (on passage in India during this period) falls prey to trappers. Birds caught by the latex method and released in urban areas do not survive after release (Ahmed 2002, Abrar Ahmed *pers. comm*. 2010). In some areas, the rollers are caught for food as well.

Conservation measures underway: This roller is protected under the Wildlife (Protection) Act, 1972. It is listed in Schedule IV of the Act. It is found in, or passes through, many PAs/IBAs. The species is recorded in a number of national monitoring schemes within its range and has been the focus of targeted study (BirdLife International 2013).

RECOMMENDATIONS

For Jammu & Kashmir, we have the following recommendations:
(1) Study its breeding ecology and habitat requirements in the Kashmir Valley.
(2) Study its movement with colour marking and satellite tracking, particularly its autumn migration and if possible spring (return) migration route.
(3) Promote organic farming and organic pesticides in the state.
(4) Strictly control all trapping, particularly during Mahashivaratri and other festivals when large numbers of rollers get caught and perish.

Long-billed Bush-warbler
Bradypterus major (Brooks 1872)

This uncommon species has a moderately small range and narrow habitat requirements, and is therefore likely to have a small global population. It is suspected to be declining as a result of habitat change, and is therefore considered Near Threatened (BirdLife International 2013).

Field Characters: A small (15 cm) olive-brown bird, with a long, thin bill, short, pale supercilium, and pale eye-ring. Chin and throat white, the throat spotted with brown, more heavily at its base (Ali & Ripley 1987), usually forming a "necklace" or gorget. As the summer progresses and the plumage becomes worn, the spots on the throat become indistinct greyish streaks and eventually disappear (Roberts 1992). Breast and vent fulvous, but belly whitish. Tail broad, slightly pointed.

Distribution: It is confined to a narrow range in western Himalaya in Pakistan and India, and in Xinjiang province of western China (Koelz 1940). In India, its main stronghold is Kashmir. It was reported from **Sonamarg** to **Baltal** in the Kashmir Valley (Bates & Lowther 1952). Pfister (2004) described Sanku in **Suru Valley** of Kargil as a typical area of encounter, confirming also the earlier sightings around **Kargil** to **Parkatchik**, and up the Indus to Upshi.

Ecology: Not much is known about the ecology and behaviour of this small bird, but that it breeds between 2,400 and 3,600 msl in low thorny scrub, rank grass, bracken, and tangled herbage on the fringes of forests (Ali & Ripley 1987). It is a great skulker and remains hidden, but during the breeding season it is easy to locate as the male sings incessantly (Roberts 1992). In winter, it is known to descend as low as 1,200 msl. According to Bates & Lowther (1952), it never enters the woods. Roberts (1992, p. 195) writes, "...it occurs in the inner Himalayan ranges on the more open slopes at the edge of forest or in terraced cultivation upto the upper limit of the tree-line...". Its preferred habitat is "the weed-grown embankments between terraced cultivation, as well as sheltered

Long-billed Bush-warbler

Himachal Pradesh

Leh

Kargil

Sonamarg

Place names
Species records

© ISRO/NRSC; [Source : www.bhuvan.nrsc.gov.in, Data : IRS-Resourcesat-1: AWIFS]

In India, most nesting records of the Long-billed Bush-warbler are from Sonamarg but it is likely to breed in more areas. Detailed study of this bird is required in the state

glades with the thickets of *Ribes grossularia* (wild gooseberry) on the edge of the spruce forest, away from cultivation…". Unlike most warblers and smaller birds which start breeding from May onwards, it appears to be a late breeder: most nests were located in July (Buchanan 1903, Osmaston 1926). Roberts (1992) reported hearing its monotonous, insect-like song from early June, when it first descends to its breeding grounds. In India, most nesting records are from Sonamarg in Kashmir. The normal clutch has four eggs, but two, and even one, have been recorded.

Threats: At present there is no obvious threat, but agricultural expansion and overgrazing (especially by goats), as well as firewood collection in the forest fringes, are likely to reduce the habitat and disturb this species in many parts of its range.

Conservation measures undertaken: It is listed in Schedule IV of the Indian Wildlife (Protection) Act, 1972, and like all Indian birds, it has some legal protection.

RECOMMENDATIONS

(1) Conduct surveys during periods of peak vocal activity in order to determine its distribution and status in Kashmir, particularly in Sonamarg and Baltal areas.

(2) Collect ecological data in order to understand its habitat requirements and identify potential threats.

(3) Protect areas of suitable habitat and safeguard against degradation.

Tytler's Leaf-warbler
Phylloscopus tytleri Brooks 1872

VINAYAK YARDI

According to BirdLife International (2013), this species has a moderately small population, which is suspected to be declining as a result of habitat loss and degradation in both the breeding and wintering grounds (perhaps more critically in the wintering grounds). It is therefore classified as Near Threatened.

Field Characters: A small (10 cm) drab olive and dull-brown bird, difficult to distinguish in the field from other similar-looking small *Phylloscopus* species. In fresh plumage, the bird is olive above and whitish below, while worn plumage becomes greyish above and dingy below (Rasmussen & Anderton 2005). Male and female are alike. A key identification point is that it has no wing bar, but it must be carefully distinguished from the very common Greenish Warbler *P. trochiloides* which sometimes has abraded plumage in mid-winter, and its wing bar is lost (Price 1980); however *P. trochiloides* has a yellower, wider beak. The similar-looking Common Chiffchaff *P. collybita* occupies more open areas, is greyer, and has conspicuously black legs. In the field in winter, one of the best ways to distinguish Tytler's Leaf-warbler is by its call, which is a long, rising *whooettt*, similar to the Willow Warbler *P. trochilus* and Common Chiffchaff, but can be differentiated from them.

Distribution: The species is endemic to the Indian subcontinent. It has a very limited breeding range in western Himalaya from northeastern Afghanistan (Nuristan) (Paludan 1959), east through Kashmir (Ali & Ripley 1987). It winters in the Western Ghats from Maharashtra to Kerala. Trevor Price observed it at Astore, north of Gilgit in Pakistan, and along the **Pangi Valley** to the west of **Udhampur** in Jammu region; these are possibly close to the northern and southern limits of

Tytler's Leaf-warbler

Himachal Pradesh

Leh

Srinagar

Udhampur

Place names
Species records

© ISRO/NRSC; [Source : www.bhuvan.nrsc-gov.in, Data : IRS-Resourcesat-1: AWiFS]

its range, respectively. The species is not known to breed on the outer slopes of the Himalaya in Himachal Pradesh. It is reported from central **Ladakh** by a single record prior to 1960 (Pfister 2004).

Ecology: Tytler's Leaf-warbler is quite well-studied, both at its breeding ground (Price 1991, Price & Jamdar 1991, Richman & Price 1992) and in wintering areas (Gross & Price 2000, Price & Gross 2005).

In summer breeding areas, it is found from 2,400 to 3,200 msl (Price 1991) where it breeds mostly in clearings and sunny margins on forest edges. It makes a small, neat nest of grass, feathers, strips of birch bark, and hair, often with a great deal of lichen in the outer part and a dense lining of feathers (Baker 1924). The nest is sometimes made on tall coniferous or birch trees, and at other times in hedgerows, from 2–10 m in height (Price & Jamdar 1991). Clutch size is 3–5, and the eggs are pure white. Although quantitative data is lacking, it is believed that only the female incubates, as in all *Phylloscopus* warblers studied so far. Food is normally small insects, gleaned warbler-like from bushes and trees. During its spring migration (March-April), many specimens have reddish pollen adhering to the forehead, and sometimes also to the chin, which may help to identify this species (Rasmussen 1998).

In winter, it forages from near ground level to the canopy (Gross & Price 2000, Price & Gross 2005). Its thin bill is correlated with its habit of foraging by picking and probing, with much less flycatching than most other leaf-warblers (Price 1991, Richman & Price 1992, Rasmussen 1998).

Threats: There is no specific threat to this tiny bird, but its wintering habitat is under tremendous anthropogenic pressure, particularly timber cutting, grazing, burning, and development. Some good forest cover has been protected in the Western Ghats.

Conservation measures undertaken: Like all other wild species in India, it is protected under the Wildlife (Protection) Act, 1972, and is listed in Schedule IV of the Act. It is present in many PAs/IBAs.

RECOMMENDATIONS

India contains most of the species' breeding range, and all of the wintering range, so this country has a major role to play in its protection. The following measures are proposed:

(1) In order to know the population trend and habitat requirements, a comprehensive study on its breeding ecology should be started in Kashmir.

(2) Similarly, another study on marked birds should be started in its wintering range, particularly in Mahabaleshwar and the Nilgiris, to determine the ideal habitat requirement and to find out whether it uses forest plantations.

(3) Annual surveys should be conducted involving trained volunteers in Kashmir to monitor its population trends.

(4) Identify the PAs/IBAs where significant numbers are seen to breed (Kashmir, Himachal), on passage (central India), and in winter (Western Ghats).

Red-headed or King Vulture
Sarcogyps calvus (Scopoli 1786)

DHRITIMAN MUKHERJEE

This vulture has suffered an extremely rapid population reduction in the recent past, which is likely to continue into the near future, probably largely as a result of feeding on carcasses of animals treated with the veterinary drug diclofenac, and perhaps in combination with other causes (BirdLife International 2013). Therefore it is listed as Near Threatened.

Distribution: This vulture, found in South Asia, has a scarce but widespread distribution in the Indian subcontinent. Recent surveys indicate (Cuthbert *et al*. 2006) that in India it has undergone a rapid population decline and is now rare or absent from some areas. Its recent distribution records from India are given by Rahmani (2012). According to Vibhu Prakash (*pers. comm*. 2013) there is no recent record from Jammu & Kashmir.

Threats and conservation measures: While there is currently no direct evidence to link the decline in this species with diclofenac poisoning, the geographic extent and rate of decline are very similar to the decline in Gyps populations for which the impact of diclofenac poisoning is now established. Counts of Red-headed Vulture carried out in 13 Protected Areas in India from 1991–1993 were repeated in 2000 and revealed a significant decline of around 48% (Prakash *et al*. 2003). The Red-headed Vulture is included in Schedule IV of the Indian Wildlife (Protection) Act, 1972. It is also listed in CITES Appendix II and CMS Appendix II.

RECOMMENDATIONS

(1) Survey Jammu regions to identify the location and number of remaining individuals and support the ban on the veterinary use of diclofenac, ketoprofen, and other similar drugs.

(2) Promote the immediate adoption of meloxicam as an alternative to diclofenac and ketoprofen.

Painted Stork
Mycteria leucocephala (Pennant 1769)

DHRITIMAN MUKHERJEE

Although one of the most abundant Asian storks, this species is classified as Near Threatened because it is thought to be undergoing a moderately rapid population decline owing to hunting, drainage, and pollution in its habitat (BirdLife International 2013), particularly in Southeast Asia.

Distribution: In India, the Painted Stork is found throughout the plains, rare in the Brahmaputra Valley and not recorded in Andaman and Nicobar Islands. In the state, it is reported only from **Gharana** Wetland in Jammu region (Singh 2013).

Ecology: The Painted Stork frequents freshwater marshes, lakes and reservoirs, flooded fields, rice paddies, freshwater swamp forest, river banks, intertidal mudflats, and salt pans. It forages in flocks in shallow waters along rivers or lakes, and nests colonially in trees, often along with other waterbirds. As it is a marginal species in the state, we are not giving its ecology in detail.

Threats: The increasing impact of habitat loss, disturbance, pollution, drainage, hunting of adults, and collection of eggs and nestlings from colonies is a cause for concern in many range countries. In India, it is protected traditionally in many areas, but poaching by tribal and amateur hunters, and pesticide poisoning are major threats. Detailed threats are mentioned by Rahmani (2012).

CONSERVATION MEASURES AND RECOMMENDATIONS

It is listed in Schedule IV of the Indian Wildlife (Protection) Act, 1972. Further, its nesting sites are traditionally protected, as a result of which its population is increasing in some areas. The wetlands of the Jammu region should be regularly monitored.

Oriental Darter
Anhinga melanogaster Pennant 1769

DHRITIMAN MUKHERJEE

The Oriental Darter is classified as Near Threatened because its population is declining moderately rapidly owing to pollution, drainage, hunting, and collection of eggs and nestlings (BirdLife International 2013).

Distribution: The species is widespread in India from coastal wetlands to about *c*. 300 m in the Himalaya. Outside India, it is found in Pakistan, Nepal, Sri Lanka, Bangladesh, Myanmar, and the whole of Southeast Asia.

Ecology: This fish-eating bird inhabits shallow inland wetlands including lakes, rivers, swamps, and reservoirs, where it hunts fish and frogs by pursuing them in clear water. It generally occurs singly or in small discrete groups, each one hunting independently. It is an expert diver and feeds almost exclusively on fish caught by its stiletto-shaped bill. It often swims with only the neck above water: the long neck and pointed bill give it the appearance of a snake, hence its popular name Snakebird. It nests colonially with egrets, storks, and herons on thorny trees, generally half-submerged or near water. Chicks are blind and naked but soon develop white down feathers which may persist even when almost fledged.

Threats: The main threat to this and all piscivorous species is from excessive fishing all over its range, particularly in India, Pakistan, and Bangladesh. Pollution and spread of invasive species such as Water Hyacinth *Eichhornea crassipes* and *Ipomea carnea* are other problems.

CONSERVATION MEASURES AND RECOMMENDATIONS

In India, it is listed in Schedule IV of the Wildlife (Protection) Act, 1972 and its hunting and disturbance are totally prohibited. Its trade is also banned. It occurs in a number of PAs/IBAs.

(1) Conduct a proper survey to determine its actual status in the state.
(2) Study its ecology and habitat requirements.

Red Kite
Milvus milvus (Linnaeus 1758)

BirdLife International (2013) considers Red Kite as Near Threatened because it is experiencing a moderately rapid population decline, owing mostly to changes in land use, poisoning from pesticides, and persecution, among other threats.

Distribution: It is vagrant in India (Ali & Ripley 1987). There are six published sight records of the Red Kite in India, mainly between January and March. In Ladakh, there is just a single record of two birds sighted over Leh (Pfister 2004).

Threats: According to BirdLife International (2013), the main threats to this species are illegal direct poisoning and indirect poisoning from pesticides and rodent bait, particularly in the wintering ranges in France and Spain, and changes in agricultural practices, causing a reduction in food resources. Illegal poisoning is believed to be preventing population growth in Scotland. Other threats include electrocution by collision with power lines, collision with wind turbines, hunting and trapping, road kills, deforestation, egg collection (on a local scale), and possibly competition with the generally more successful Black Kite *M. migrans*. Another factor implicated in the decline in France and Spain is a decrease in the number of rubbish dumps (BirdLife International 2013).

CONSERVATION MEASURES AND RECOMMENDATIONS

The Red Kite is protected under the Wildlife Protection Act, 1972, and listed in Schedule I. As this bird is vagrant in India, we do not have any specific recommendations for it. India is way beyond its normal distribution range and surveys would not yield much information.

Pallid Harrier
Circus macrourus (Gmelin 1770)

DHRITIMAN MUKHERJEE

According to BirdLife International (2013), the Pallid Harrier is known to be undergoing steep population decline in Europe, although the numbers in its Asiatic strongholds are thought to be more stable. Thus, it is probably experiencing a moderately rapid population decline overall, and consequently it is categorised as Near Threatened.

Distribution: The Pallid Harrier is a winter visitor to the entire Indian subcontinent, affecting grasslands, undulating foothills with grass, open scrub, and crop fields. It avoids forest.

Ecology: In winter, it is found in open countryside, either singly or in small groups, systematically skirting the ground for prey. It is seen from sea level to 3,000 msl in the Himalaya and up to 4,000 msl in Africa. It is usually silent in winter, but occasionally heard calling *keck-keck-keck-keck* at dusk before finally settling to roost (Naoroji 2007). It breeds in wet grasslands close to small rivers and lakes, marshlands, semi-desert, steppe and forest-steppe, and sometimes even in boreal and tundra forest zones, from sea level to 1,200 msl. Clutch size is four to five and incubation up to 30 days. Incubation is done by the female alone, with the male bringing food to her during incubation and early brooding.

Threats: The main threat in the breeding areas is the destruction and degradation of grasslands through conversion of arable land to agriculture, burning of vegetation, intensive grazing of wet pastures, and clearance of shrubs and tall weeds.

CONSERVATION MEASURES AND RECOMMENDATIONS

Like all raptors, it is also protected under the Indian Wildlife (Protection) Act, 1972, and listed in Schedule I of the Act. Scientific survey should be done to find out its distribution in Jammu & Kashmir.

Little Bustard
Tetrax tetrax (Linnaeus 1758)

BirdLife International (2013) justifies its status as Near Threatened because the Little Bustard is probably experiencing a moderately rapid overall population decline, driven by rapid decline in the west of its range, owing mainly to habitat loss and degradation, as well as low-level hunting pressure. It is a marginal species in the state of Jammu & Kashmir.

Distribution: The Little Bustard was a very rare winter migrant to undivided India, mostly shot in Peshawar and adjoining districts (now in Pakistan). In India, most records were from Kashmir. Ali & Ripley (1987) have mentioned records from Ludhiana and Gurdaspur district in Punjab, and Saharanpur in Uttar Pradesh, but Rasmussen & Anderton (2005) state that no specimen has been traced east of Kashmir. The latest record is from **Haigam**, Kashmir in January, 1964 by Col. H. Nedou (Ali & Ripley 1987). There is no record of the Little Bustard for the last 50 years in India, so it is not being described in detail here.

Ecology: This species is omnivorous, consuming seeds, insects, rodents, and reptiles. Like other bustards, the male Little Bustard has a flamboyant courtship display, with foot-stamping and leaping in the air. Females lay three to five eggs on the ground. The habitat is open grassland and undisturbed cultivation, with plants tall enough for cover. It has a slow, stately walk, and when disturbed tends to run rather than fly. It is gregarious, especially in winter.

Threats: The primary cause of its decline has been conversion of dry grassland and low-intensity cultivation to intensive agriculture, especially where this has included the planting of monoculture or perennial crops, irrigation or afforestation (BirdLife International 2013).

CONSERVATION MEASURES AND RECOMMENDATIONS

Like all bustard species, the Little Bustard is also protected under the Indian Wildlife (Protection) Act, 1972. It is listed in Schedule IV of the Act. It is also included in CITES Appendix II. A European action plan was published in 2001. As it is a marginal species with no recent record, we do not make any specific recommendations.

REFERENCES

Ahmed, A. (2002) Live Bird Trade in India. Unpublished report. TRAFFIC India/WWF-India, New Delhi.

Akhtar, A., Prakash, V. and Javed, S. (1994) The Western Tragopan – bird of the Himalaya. *Sanctuary (Asia)* 14(2): 44–49.

Akhtar, S.A. (1989) Some Observations on the Breeding Behaviour of the Black-necked Crane (*Grus nigricollis*) in Ladakh. Asia Crane Congress, Rajkot, India. Pp. 17.

Ali, S. and Ripley, S.D. (1987) *Compact Handbook of the Birds of India and Pakistan together with those of Bangladesh, Nepal, Bhutan and Sri Lanka.* 2nd edn. Oxford University Press, Delhi. Pp. 890.

Anon. (2004) Report on the International South Asian Vulture Recovery Plan Workshop. *Buceros* 9(1): 48.

Baker, E.C.S. (1908) *Indian Ducks and Their Allies*. Bombay Natural History Society, Bombay.

Baker, E.C.S. (1921) *Indian Ducks and their Allies.* 2nd edn. Bombay Natural History Society, Bombay.

Baker, E.C.S. (1922) *The Fauna of British India, including Ceylon and Burma. Birds*. Vol. I. 2nd edn. Taylor and Francis, London.

Baker, E.C.S. (1924) *The Fauna of British India, including Ceylon and Burma. Birds*. Vol. II. 2nd edn. Taylor and Francis, London.

Baral, H.S., Giri, J.B. and Virani, M.Z. (2005) On the decline of Oriental Whitebacked Vultures *Gyps bengalensis* in lowland Nepal. Pp. 215–19. In: Chancellor, R.D. and Meyburg, B.-U. (eds) *Raptors Worldwide. Proceedings of the 6th World Conference on Birds of Prey and Owls.* WWGBP, Berlin and MME/Birdlife Hungary, Budapest.

Barter, M., Cao, L., Chen, L. and Lei, G. (2005) Result of a survey for waterbirds in the lower Yangtze floodplains, China in January-February 2004. *Forktail* 21: 1–7.

Bates, R.S.P. and Lowther, E.N.H. (1952) *Breeding Birds of Kashmir*. Oxford University Press, Delhi. Pp xxiii+369.

Bean, N.J., Benstead, P.J., Showler, D.A. and Whittington, P.A. (1994) Survey of Western Tragopan *Tragopan melanocephalus* in the Palas Valley, NWFP – spring 1994. Unpublished.

Betts, F.N. (1954) Occurrence of the Blacknecked Crane (*Grus nigricollis*) in Indian limits. *J. Bombay Nat. Hist. Soc.* 52: 605–606.

Bhargava, R. (2001) Record of Yellow-eyed Pigeon *Columba eversmanni* from Meerut district of Uttar Pradesh, India. *OBC Bulletin* 34: 36–37.

Bhatt, K. (2006) Black-necked Stork *Ephippiorhynchus asiaticus* nest with four chicks in Marine National Park, Gujarat, India. *Indian Birds* 2(2): 35.

BirdLife International (2001) *Threatened birds of Asia: the Birdlife International Red Data Book*. Eds: Collar, N.J., Andreev, A.V., Chan, S., Crosby, M.J., Subramanya, S., and Tobias, J.A. Birdlife International, Cambridge, UK.

BirdLife International (2013) Species factsheets. Downloaded from http://www.birdlife.org.

Bisht, M.S., Phurailatpam, S., Kathait, B.S., Dobriyal, A.K., Chandola-Saklani, A. and Kaul, R. (2007) Survey of threatened Cheer Pheasant *Catreus wallichii* in Garhwal Himalaya. *J. Bombay Nat. Hist. Soc.* 104(2): 134–39.

Buchanan, K. (1903) Nesting Notes from Kashmir. *J. Bombay Nat. Hist. Soc.* 15: 131–33.

Chacko, R.T. (1992a) Black-necked cranes wintering in Bhutan. *OBC Bull.* 16: 36–38.

Chacko, R.T. (1992b) Black-necked Cranes in Bhutan: a full winter study, October 1991–April 1992. Unpublished Report, Oriental Bird Club and Bhutan Dept. Forestry, Bangalore, India. Pp. 33.

Chacko, R.T. (1993a) Blacknecked Cranes wintering in Bhutan. *Newsletter for Birdwatchers* 33(2): 23–25.

Chacko, R.T. (1993b) Human interference in the habitats of cranes in Bhutan and Ladakh. *Newsletter for Birdwatchers* 33: 106–108.

Chacko, R.T. (1995) A Summer 95 Study of the Black-necked Cranes Breeding in Some Remote High Altitude Areas of Ladakh, India. Unpublished Report.

Chacko, R.T. (1996) A Summer 96 Study of the Black-necked Cranes Breeding in Some Remote High Altitude Areas of Ladakh, India. Unpublished Report.

Chandan P., Chatterjee, A., Gautam, P., Seth, C.M., Takpa, J., Haq, S., Tashi, P. and Vidya, S. (2005) Black-necked Crane – Status, Breeding Productivity and Conservation in Ladakh, India 2000–2004. WWF-India and Department of Wildlife Protection, Government of Jammu & Kashmir.

Choudhury, A.U. (1996) The Black-necked Crane in Arunachal Pradesh. *The Twilight* 2 (2&3): 31–32.

Choudhury, A.U. (2002) Status and Conservation of Cranes in Northeast India. Pp. 41–44. In: Rahmani, A.R. and Gayatri Ugra (eds) *Birds of Wetlands and Grassland: Proceedings of the Sálim Ali Centenary Seminar on Conservation of Avifauna of Wetlands and Grasslands.* Pp. x + 228. Bombay Natural History Society, Mumbai.

Cramp, S. & Simmons, K.E.L. (eds) (1977) *Handbook of the Birds of Europe, the Middle East and North Africa.* Vol. 1: Ostrich to Ducks. Oxford University Press, Oxford, London, and New York.

Cuthbert, R., Green, R.E., Ranade, S., Saravanan, S., Pain, D.J., Prakash, V. and Cunningham, A.A. (2006) Rapid population declines of Egyptian Vulture (*Neophron percnopterus*) and Red-headed Vulture (*Sarcogyps calvus*) in India. *Animal Conservation* 9: 349–354.

Cuthbert, R., Taggart, M.A., Prakash, V., Saini, M., Swarup, D., Upreti, S., Mateo, R., Chakraborty, S.S., Deori, P. and Green, R.E. (2011) Effectiveness of action in India to reduce exposure of *Gyps* vultures to the toxic veterinary drug Diclofenac. PLoS ONE 6(5): e19069. doi:10.1371/journal.pone.0019069.

Daniel, J.C., Balachandran, S. and Alagarrajan, S. (1999) *Community participation in conservation of the waterbirds of Vedaranyam Swamp. A case study on bird trappers.* Sálim Ali Wild Wings Trust, Mumbai.

del Hoyo, J., Elliot, A. and Sargatal, J. (eds) (1992) *Handbook of the Birds of the World. Vol. 1: Ostriches to Ducks.* Lynx Edicions, Barcelona, Spain.

del Hoyo, J., Elliott, A. and Sargatal, J. (1996) *Handbook of the Birds of the World, Vol. 3: Hoatzin to Auks.* Lynx Edicions, Barcelona, Spain.

del Hoyo, J., Elliott, A., and Sargatal, J. (eds) (1997) *Handbook of the Birds of the World. Vol. 4: Sandgrouse to Cuckoos.* Lynx Edicions, Barcelona.

del Hoyo, J., Elliott, A. and Sargatal, J. (eds) (2001) *Handbook of the Birds of the World. Vol. 6: Mousebirds to Hornbills.* Lynx Edicions, Barcelona.

Delacour, J. (1951) *Pheasants of the World.* Country Life Limited, London and Charles Scribner's Sons, New York. Pp. 351.

Ferguson-Lees, J. and Christie, D.A (2001) *Raptors of the World.* Christopher Helm, London.

Fuller, R.A. and Garson, P.J. (2000) *Pheasants: status survey and conservation action plan 2000–2004.* WPA/BirdLife/SSC Pheasant Specialist Group, IUCN, Gland, Switzerland, and the World Pheasant Association, Reading, UK.

Garson, P.J., Young, L. and Kaul, R. (1992) Ecology and Conservation of the Cheer Pheasant *Catreus wallichii:* Status in the wild and the progress of a reintroduction project. *Biol. Conserv.* 59: 25–35.

Gaston, A.J., Islam, K. and Crawford, J.A. (1983) The current status of the Western Tragopan *Tragopan melanocephalus. J. World Pheasant Assoc.* 8: 40–49.

Ghosh, S. (1997) Record of Chir Pheasant, *Catreus wallichii* above 4,545 metres in the Western Himalayas. *J. Bombay Nat. Hist. Soc.* 94(3): 566.

Gilbert, M., Watson, R.T., Virani, M.Z., Oaks, J.L., Ahmed, S., Chaudhry, M.J.I., Arshad, M. and Mahmood, S. (2006) Rapid population declines and mortality clusters in three Oriental white-backed vulture *Gyps bengalensis* colonies in Pakistan due to diclofenac poisoning. *Oryx* 40: 388–399.

Gole, P. (1981) Black-necked Cranes in Ladakh. Pp. 197–203. *In:* Lewis J.C. and H. Masotomi (eds) *Crane research around the world: Proceedings of the International Crane Symposium at Sapporo, Japan in 1980 and papers from the World Working Group on Cranes, International Council for Bird Preservation.* International Crane Foundation, Baraboo, USA.

Gole, P. (1983) Future of Black-necked Cranes in Indian Sub-continent. Pp. 51–54. In: Archibald, G.W. and Pasquier, R.F. (eds) *Proceedings of the International Crane Workshop Bharatpur, India.* International Crane Foundation, Baraboo, Wisconsin, USA.

Gole, P. (1996) *A Guide to the Cranes of India.* Bombay Natural History Society, Bombay. Pp. 35.

Green, A. and Anstey, S. (1992) The status of the White-headed Duck *Oxyura leucocephala. Bird Conservation International* 2: 185–200.

Green, A.J. (2000) The habitat requirements of the Marbled Teal (*Marmaronetta angustirostris* Ménétr.) A review. Pp. 131–140. In: Comín, F.A., Herrera, J.A. and Ramírez, J. (eds) *Limnology and Aquatic Birds: Monitoring, Modelling and Management.* Universidad Autónoma del Yucatán, Mérida.

Green, A.J. and Hughes, B. (1996) Action plan for the White-headed Duck (*Oxyura leucocephala*). Pp. 119–145. In: Heredia, B., Rose, L. and Painter, M. (eds) *Globally Threatened Birds in Europe: Action Plans.* Council of Europe, Strasbourg.

Green, A.J., Fox, A.D., Hughes, B. and Hilton, G.M. (1999) Time-activity budgets and site selection of White-headed Duck *Oxyura leucophala* at Burdur Lake, Turkey in late winter. *Bird Study* 46: 62–73.

Green R.E., Newton, I., Shultz, S., Cunningham, A.A., Gilbet, M., Pain, D.J. and Prakash, V. (2004) Diclofenac poisoning as a cause of vulture population declines across the Indian subcontinent. *J. Anim. Ecol.* 41: 793–800.

Grimmett, R., Inskipp, T. and Inskipp, C. (1998) *Birds of the Indian Subcontinent.* Christopher Helm, London.

Gross, S. and Price, T. (2000) Determinants of the northern and southern range limits of a warbler. *Journal of Biogeography* 27: 869–878.

Henry, G.M. (1955) *A Guide to the Birds of Ceylon.* Oxford University Press, Colombo and London. Pp. 432.

Hussain, S.A. (1976) Preliminary Report. Bombay Natural History Society/World Wildlife Fund – India. Ladakh Expedition.

Hussain, S.A. (1985) Status of Black-necked Crane in Ladakh – 1983, Problems and Prospects. *J. Bombay Nat. Hist. Soc.* 82: 449–58.

Ishtiaq, F. (1998) Comparative ecology and behaviour of storks in Keoladeo National Park, Rajasthan, India. Ph.D. Thesis. Aligarh Muslim University, Aligarh, India.

Ishtiaq, F., Javed, S., Coulter, M.C. and Rahmani, A.R. (2010) Resource partitioning in three sympatric species of storks in Keoladeo National Park, India. *Waterbirds* 33(1): 41–49.

Ishtiaq, F., Rahmani, A.R., Javed, S. and Coulter, M.C. (2004) Nest-site characteristics of Black-necked Stork *Ephippiorhynchus asiaticus* and Woolly-necked Stork *Ciconia episcopus* in Keoladeo National Park, Bharatpur, India. *J. Bombay Nat. Hist. Soc.* 101(1): 90–95.

Islam, K. and Crawford, J.A. (1987) Habitat use by Western Tragopan *Tragopan melanocephalus* (Gray) in north-eastern Pakistan. *Bio. Conserv.* 40: 101–115.

Islam, M.Z. and Rahmani, A.R. (2004) *Important Bird Areas in India: Priority sites for conservation.* Indian Bird Conservation Network, Bombay Natural History Society, and BirdLife International, UK. Pp. xviii + 1133.

Jandrotia, J.S., Katoch, S.S., Kaul, R. and Seth, K. (2000) Surveys of pheasants in Chamba District, Himachal Pradesh, India. Pp. 75–78. In: Woodburn, M., McGowan, P., Carroll, J., Musavi, A. and Zhen-wang, Z. (eds) *Galliformes 2000 – Proceedings of the 2nd International Galliformes Symposium.* World Pheasant Association, Reading, UK.

Jandrotia, J.S., Sharma, V. and Katoch, S.S. (1995) A pheasant survey in the Ravi Catchment of Chamba District, Himachal Pradesh, India. *Ann. Rev. WPA 1994/95:* 67–74.

Javed, S. (1992) Birds of Limber valley forest (Jammu and Kashmir). *Newsletter for Birdwatchers* 32(5/6): 13–15.

Johnsgard, P.A. and Carbonell, M. (1996) *Ruddy Ducks and other stifftails, their biology and behaviour.* Univ. Oklahoma Press, Norman.

Johnson, J.A., Lerner, Heather R.L., Rasmussen, P.C. and Mindell, D.P. (2006) Systematics within Gyps vultures: a clade at risk. *BMC Evolutionary Biology* 6: 65 doi:10.1186/1471-2148-6-65.

Kaul, R. (1989) Western Tragopan surveys in the Limber valley, Kashmir, India. *WPA News* 26: 12–14.

Kaur, J. and Choudhury, B.C. (2003) Stealing of Sarus Crane eggs. *Curr. Sci.* 85(11): 1515–1516.

Kaur, J. and Nair, A. (2008) Community involvement in conservation of Sarus Crane breeding habitat in three districts of semi-arid tract of Rajasthan, India. Report submitted to Rufford Small Grants Foundation, UK.

Knox, A.G. and Walters, M.P. (1994) Extinct and endangered birds in the collections of the Natural History Museum. British Ornithologists' Club, London.

Koelz, W. (1940) Notes on the birds of Zanskar and Purig, with appendices giving new records for Ladakh, Rupshu and Kulu. *Pap. Michigan Acad. Sci. Arts Letters* 25: 297–322.

Kylin, H. (2002) Kashmir Flycatcher *Ficedula subrubra* nesting in Sri Lanka? *Bull. Oriental Bird Club* 36: 73.

Lawrence, W.R. (1895) *The Valley of Kashmir*. H. Frowde, London.

Ludlow, F. (1920) Notes on the nidification of certain birds in Ladak. *J. Bombay Nat. Hist. Soc.* 27: 141–146.

Madsen, J. (1996) International action plan for the Lesser White-fronted Goose *Anser erythropus*. Pp. 67–78. In: Heredia, B., Rose, L. and Painter, M. (eds) *Globally threatened birds in Europe: Action plans*. Council of Europe, Germany.

Maheswaran, G. (1998) Ecology and behaviour of the Black-necked Stork *Ephippiorhynchus asiaticus* in Dudhwa National Park, Uttar Pradesh, India. Ph.D. Thesis. Aligarh Muslim University, Aligarh, India.

Meinertzhagen, R. (1927) Systematic Results of Birds Collected at High altitudes in Ladakh and Sikkim, Part 11. *Ibis* 69(3): 571–633.

Mukherjee, A., Borad, C.K., Parasharya, B.M. and Soni, V.C. (2001) Factors affecting distribution of the Sarus Crane *Grus antigone antigone* (Linn.) in Kheda District, Gujarat. *J. Bombay Nat. Hist. Soc.* 98(3): 379–384.

Muralidharan, S. (1992) Poisoning of Sarus. *Hornbill* 1992(1): 3–7.

Naoroji, R. (2007) *Birds of Prey of the Indian Subcontinent*. Om Books International, New Delhi.

Naoroji, R. and Sangha, H. (2003) Project Golden Eagle: Raptor surveys in Ladakh 1997–2003. Unpublished Report.

Narayan, G., Akhtar, A., Rosalind, L. and D'Cunha, E. (1987) Blacknecked Crane (*Grus nigricollis*) in Ladakh. 1986. *J. Bombay Nat. Hist. Soc.* 83(4): 180–195.

Nurbu, C. (1983) Notes on the Black-necked Crane in Ladakh. Pp. 55–56. *In:* Archibald, G.W. and R.F. Pasquier (eds) *Proceedings of the 1983 International Crane Workshop, Bharatpur India*. International Crane Foundation, Baraboo, Wisconsin, USA.

Oaks, J.L., Gilbert, M., Virani, M.Z., Watson, R.T., Meteyer, C.U., Rideout, B., Shivaprasad, H.L., Ahmed, S., Chaudhry, M.J.I., Arshad, M., Mahmood, S., Ali, A. and Khan, A.A. (2004a) Diclofenac residues as the cause of vulture population decline in Pakistan. *Nature* 427: 630–633.

Oaks, J.L., Donahoe, S.L., Rurangirwa, F.R., Rideout, B.A., Gilbert, M., Virani, M.Z. (2004b) Identification of a Novel Mycoplasma Species from an Oriental White-Backed Vulture (*Gyps bengalensis*) . *J. Clin. Microbiol.* 42: 5909–5912.

Oliver, D.G. (1919) Spot Bill Duck in Kashmir. *J. Bombay Nat. Hist. Soc.* 26: 675.

Osmaston, B.B. (1925) On the Birds of Ladakh. *Ibis* 12: 662.

Osmaston, B.B. (1926) Birds nesting in the Dras and Suru Valleys. *J. Bombay Nat. Hist. Soc.* 31: 186–197.

Pain, D.J., Bowden, C.G.R., Cunningham, A.A., Cuthbert, R., Das, D., Gilbert, M., Jakati, R.D., Jhala, Y., Khan, A.A., Naidoo, V., Oaks, J.L., Parry-Jones, J., Prakash, V., Rahmani, A., Ranade, S.P., Baral, H.S., Senacha, K.R. and Saravanan, S. (2008) The race to prevent the extinction of South Asian vultures. *Bird Conservation International* 18: 30–48.

Paludan, K. (1959) On the birds of Afghanistan. *Vidensk. Medd. Dansk Naturhist. Foren.* 122: 1–332.

Pandey, S. (1995) A preliminary estimate of numbers of Western Tragopan in Daranghati Sanctuary, Himachal Pradesh. *Ann. Rev. WPA 1993/94*: 49–56.

Pfister, O. (1998) The breeding ecology and conservation of the Black-necked Crane (*Grus nigricollis*) in Ladakh, India. Unpublished Thesis. University of Hull, Hull, UK. Pp. 136.

Pfister, O. (2004) *Birds and Mammals of Ladakh*. Oxford University Press, New Delhi.

Prakash, V. (1999) Status of vultures in Keoladeo National Park, Bharatpur, Rajasthan, with special reference to population crash in *Gyps* species. *J. Bombay Nat. Hist. Soc.* 96: 365–378.

Prakash, V. and Nanjappa, C. (1988) An instance of active predation by Scavenger Vulture (*Neophron p. ginginianus*) on Checkered Keelback Watersnake (*Xenochrophis piscator*) in Keoladeo National Park, Bharatpur, Rajasthan. *J. Bombay Nat. Hist. Soc.* 85(2): 419.

Prakash, V., Green, R.E., Pain, D.J., Ranade, S.P., Saravanan, S., Prakash, N., Venkitachalam, R., Cuthbert, R., Rahmani, A.R. and Cunningham, A.A. (2007) Recent changes in populations of resident *Gyps* vultures in India. *J. Bombay Nat. Hist. Soc.* 104: 129–135.

Prakash, V., Pain, D.J., Cunningham, A.A., Donald, P.F., Prakash, N., Verma, A., Gargi, R., Sivakumar, S. and Rahmani, A.R. (2003) Catastrophic collapse of Indian White-backed *Gyps bengalensis* and Long-billed *Gyps indicus* vulture populations. *Biological Conservation* 109 (3): 381–390.

Price, T.D. (1980) On the occurrence of Tytler's Leaf Warbler *Phylloscopus tytleri* Brooks in Goa. *J. Bombay Nat. Hist. Soc.* 77: 143–144.

Price, T.D. (1991) Morphology and ecology of breeding warblers along an altitudinal gradient in Kashmir, India. *J. Anim. Ecol.* 60: 643–664.

Price, T.D. and Gross, S. (2005) Correlated evolution of ecological differences among the Old World Leaf Warblers in the breeding and non-breeding seasons. Pp. 359–372. In: Greenberg, R. and Marra, P. (eds) *Birds of Two Worlds*. Smithsonian Institution Press, Washington, DC.

Price, T.D. and Jamdar, N. (1990) The breeding birds of Overa Wildlife Sanctuary, Kashmir. *J. Bombay Nat. Hist. Soc.* 87(1): 1–15.

Price, T.D. and Jamdar, N. (1991) Breeding of eight sympatric species of *Phylloscopus* warblers in Kashmir. *J. Bombay Nat. Hist. Soc.* 88: 242–255.

Price, T.D., Zee, J., Jamdar, K. and Jamdar, N. (2003) Bird species diversity along the Himalayas: a comparison of Himachal Pradesh with Kashmir. *J. Bombay Nat. Hist. Soc.* 100(2&3): 394–409.

Qadri, M.Y., Kaul, R. and Iqbal, M. (1990) Status of pheasants in Kashmir with special reference to endangered species. Pp. 124–128. In: Hill, D.A., Garson, P.J. and Jenkins, D. (eds) *Pheasants in Asia 1989*. World Pheasant Association, Reading, UK.

Qian, F., Wu, H., Gao, L., Zhang, H., Li, F., Zhong, H., Yang, X. and Zheng, G. (2009) Migration routes and stopover sites of Black-necked Cranes determined by satellite tracking. *J. Field Ornithol.* 80(1): 19–26.

Rahmani, A.R. (1989) Status of the Black-necked Stork *Ephippiorhynchus asiaticus* in the Indian subcontinent. *Forktail* 5: 99–110.

Rahmani, A.R. (2008) Race to save Vultures. *Hornbill* Oct–Dec 2008: 147–55.

Rahmani, A.R. (2012) *Threatened Birds of India – Their Conservation Requirements*. Indian Bird Conservation Network: Bombay Natural History Society, Royal Society for the Protection of Birds and BirdLife International. Oxford University Press. Pp. xvi + 864.

Rahmani, A.R. and Islam, M.Z. (2008) *Ducks, Geese and Swans of India*. Indian Bird Conservation Network, Bombay Natural History Society, Royal Society for the Protection of Birds and BirdLife International. Oxford University Press, Delhi. Pp. 374.

Ramachandran, N.K. and Vijayan, V.S. (1994) Distribution and general ecology of the Sarus Crane (*Grus antigone*) in Keoladeo National Park, Bharatpur, Rajasthan. *J. Bombay Nat. Hist. Soc.* 91(2): 210–223.

Ramesh, K. (2003) An ecological study on the pheasants of Great Himalayan National Park, Western Himalaya. Ph.D. Thesis, Forest Research Institute, Dehra Dun, India.

Ramesh, K., Sathyakumar, S. and Rawat, G.S. (2008) Methods of capture and radio tracking of Western Tragopan *Tragopan melanocephalus* J.E. Gray 1829 in the Great Himalayan National Park, India. *J. Bombay Nat. Hist. Soc.* 105(2): 127–132.

Rana, G. and Prakash, V. (2004) Unusually high mortality of cranes in areas adjoining Keoladeo National Park, Bharatpur, Rajasthan. *J. Bombay Nat. Hist. Soc.* 101(2): 317.

Rasmussen, P.C. (1998) Tytler's Leaf Warbler *Phylloscopus tytleri*: non-breeding distribution, morphological discrimination, and ageing. *Forktail* 14 (August): 17–28.

Rasmussen, P.C. and Anderton, J.C. (2005) *Bird of South Asia: The Ripley Guide*. Vols 1&2. National Museum of Natural History – Smithsonian Institution, Michigan State University and Lynx Edicions, Washington D.C. and Barcelona.

Richman, A.D. and Price, T.D. (1992) Evolution of ecological differences in the Old World leaf warblers. *Nature* 355: 817–821.

Roberts, T.J. (1991) *The Birds of Pakistan*. Vol. 1. Non-Passeriformes. Oxford University Press, Karachi.

Roberts, T.J. (1992) *The Birds of Pakistan*. Vol. 2. Passerines. Oxford University Press, Karachi.

Sahi, D.N. (1993) Wildlife conservation sites in Kashmir Himalayas. *Tigerpaper* 20(2): 28–31.

Sanchez, M.I., Green, A.J. and Dolz, C. (2000) The diets of the White-headed Duck *Oxyura leucocephala,* Ruddy Duck *O. jamaicensis* and their hybrids from Spain. *Bird Study* 47: 275–84.

Seibold, I. and Helbig, A.J. (1995) Evolutionary history of New and Old World vultures inferred from nucleotide sequence of the mitochondrial cytochrome b gene. *Phil. Trans. R. Soc. London* B, 350: 163–178.

Senacha, K.R., Taggart, M.A., Rahmani, A.R., Jhala, Y.V., Cuthbert, R., Pain, D.J. and Green, R.E. (2008) Diclofenac levels in livestocks carcasses in India before the 2006 "ban". *J. Bombay Nat. Hist. Soc.* 105(2): 148–61.

Shultz, S., Baral, H.S., Charman, S., Cunningham, A.A., Das, D., Ghalsasi, D.R., Goudar, M.S., Green, R.E., Jones, A., Nighot, P., Pain, D.J. and Prakash, V. (2004) Diclofenac poisoning is widespread in declining vulture populations across the Indian subcontinent. *Proceedings of the Royal Society of London*, B (Supplement), 271: 458–460.

Singh, R. (2013) *Plumage Across the Pir Panjal: The Poonch and Rajouri Districts.* Samrat Offset Pvt. Ltd., New Delhi.

Sundar, K.S.G. (2003) Notes on the breeding biology of the Black-necked Stork *Ephippiorhynchus asiaticus* in Etawah and Mainpuri districts, Uttar Pradesh, India. *Forktail* 19: 15–20.

Sundar, K.S.G. (2009) Are rice paddies suboptimal breeding habitat for Sarus Cranes in Uttar Pradesh, India? *The Condor* 111: 611–623.

Sundar, K.S.G. and Choudhury, B.C. (2001) A note on Sarus Crane *Grus antigone* mortality due to collision with high-tension power lines. *J. Bombay Nat. Hist. Soc.* 98(1): 108–110.

Sundar, K.S.G. and Choudhury, B.C. (2003) The Indian Sarus Crane *Grus a. antigone*: a literature review. *Journal of the Ecological Society* 16: 16–41.

Sundar, K.S.G. and Choudhury, B.C. (2006) Conservation of the Sarus Crane *Grus antigone* in Uttar Pradesh, India. *J. Bombay Nat. Hist. Soc.* 103 (2&3): 182–190.

Sundar, K.S.G., Deomurari, A., Bhatia, Y. and Narayanan, S.P. (2007) Records of Black-necked Stork *Ephippiorhynchus asiaticus* breeding pairs fledging four chicks. *Forktail* 23: 161–163.

Sundar, K.S.G., Kaur, J. and Choudhury, B.C. (2000) Distribution, demography and conservation status of the Indian Sarus Crane (*Grus antigone antigone*) in India. *J. Bombay Nat. Hist. Soc.* 97(3): 319–39.

Swan, G., Naidoo, V., Cuthbert, R., Green, R.E., Pain, D.J., Swarup, D., Prakash, V., Taggart, M., Bekker, L., Das, D., Diekmann, J., Diekmann, M., Killian, E., Meharg, A., Patra, R.C., Saini, M. and Wolter, K. (2006a) Removing the threat of diclofenac to critically endangered Asian vultures. *PLoS Biol.* 4(3): e66.doi: 10,1371/journal. pbio.0040066.

Swan, G.E., Cuthbert, R., Quevedo, M., Green, R.E., Pain, D.J., Bartels, P., Cunningham, A.A., Duncan, N., Meharg, A.A., Oaks, J.L., Parry-Jones, J., Taggart, M.A., Verdoorn, G. and Wolter, K. (2006b) Toxicity of diclofenac to *Gyps* vultures. *Biol. Lett.* 279–82.

Unwin, W.A. (1897) Late stay of wildfowl. *J. Bombay Nat. Hist. Soc.* 11: 169.

Ward, A.E. (1906–1908) Birds of the provinces of Kashmir and Jammu and adjacent districts. *J. Bombay Nat. Hist. Soc.* 17: 108–113, 479–485, 723–729, 943–949; 18: 461–464.

Williams, C. and Delany, S. (1996) Migration through the north-west Himalaya – Some Results of the Southampton University Ladakh Expeditions. Part II. *OBC Bulletin* 3: 11–16.

Wright, R.G. and Dewar, D. (1925) *The Ducks of India*. Witherby, London.

Zarri, A.A. and Rahmani, A.R. (2004) Wintering records, ecology and behaviour of Kashmir Flycatcher *Ficedula subrubra* (Hartert and Steinbacher). *J. Bombay Nat. Hist. Soc.* 101(2): 261–268.

■ ■ ■

Index of Common Names

Index of Scientific Names

Dato Loke Wan Tho – a Dedication

Dato Loke Wan Tho was a businessman, ornithologist, photographer, philanthropist, and a great supporter of BNHS.

In his lifetime, Loke Wan Tho was honoured by the state of Kelantan in Malaysia from whom he received his Datoship, Cambodia, Japan, and Malaya. Always the philanthropist, he supported many charities, associations, and educational institutions.

Dato Loke Wan Tho was an unusual man for his time – a Renaissance Man in a world that was once the backwater of an Empire – he was an ornithologist and lover of nature long before it became popular to care for the environment; he was a man who kept his word, a sportsman who played for the love of the sport despite his poor health, and he was a collector of art in its many forms whilst many did not know how to appreciate it.

His greatest legacy, however, was to his family: he set a priceless example to many, and today they owe their love of fauna and flora, their various interests in sport and academia, their passion for collecting art, and their contributions to society to his direct influence. He was a man of many interests that ranged from books to collecting art to golf. He was very supportive of local art and sponsored many artists.

He played a good game of golf and became the first President of the Singapore Island Country Club after it was merged. However his greatest passion was photography and ornithology. This passion took him on expeditions to Papua New Guinea to photograph the Birds of Paradise; Cambodia where a firm and lasting friendship with King Sihanouk was forged; India, the Hindu Kush, Kashmir, Sikkim and the wild yonder, always with his camera and Dr. Sálim Ali for company.

As one writer said of him, "Generous to a fault, whether he gave to State, Charity or Institution, to aspiring artist or struggling student, his gifts were made after careful consideration and always with the minimum of ostentation. Indeed, it was his tact, conscientiousness and complete integrity and efficiency with which he discharged all his functions which won him the respect and admiration of all."

From the autobiography of Dr. Sálim Ali, *The Fall of a Sparrow*, we are able to glean a vignette from the diary of his friend. About the Thangu region, Loke Wan Tho records: "Everywhere one sees prayer flags. The prayers are printed on them from blocks which may be had at the monasteries. Do we, who believe in Science, believe in all this, or do we say with Tolstoy 'Science, that is the supposed knowledge of absolute truth', or guardedly pray with the Scientist 'O Lord, if there be one, save my soul, if any'." This one statement speaks volumes about Loke Wan Tho, his scientific bent of mind, keen observation, sense of humour, and most of all his unique way with words.

We at the Bombay Natural History Society are privileged to hold a part of his legacy, a fabulous collection of bird photographs, and numerous landscapes of the remote and pristine areas that he explored while accompanying Dr. Sálim Ali on his field trips. The pictures bear visual testimony to the excellence of the man behind the camera.

Compiled from Wikipedia

Some important publications by Loke Wan Tho

Loke, W.T. (1945) Strange death of a young Cuckoo (*Cuculus canorus*). *JBNHS* 45: 419–420.

Loke, W.T. (1945) Notes on the behaviour of nesting Paddy Birds (*Ardeola grayii*) in Kashmir. *JBNHS* 45: 608–609.

Loke, W.T. (1946) A bird photographer in Kashmir. *JBNHS* 46: 431–436.

Loke, W.T. (1952) Photographing the Whitebellied Sea-Eagle (*Haliaetus leucogaster* (Gmelin)). *JBNHS* 50: 618–622.

Loke, W.T. (1952) Photographing the Whitebellied Sea-Eagle (*Haliaetus leucogaster* (Gmelin)). JBNHS 50: 658.

Loke, W.T. (1952) The orthography of English names of birds. *JBNHS* 50: 678–679.

Loke, W.T. (1952) Photographing birds with the highspeed flash. *JBNHS* 50: 785–786.

Loke, W.T. (1952) Kashmir revisited. *JBNHS* 51: 121–127.

Loke, W.T. (1956) A Dabchick is born. *JBNHS* 53: 468–470.

Loke, W.T. (1956) Experiences with Little Ringed Plover. *JBNHS* 54: 185–188.

Loke, W.T. (1957) *A Company of Birds*. Michael Joseph, London.

Loke, W.T. (2004) Photographing birds with the highspeed flash *Hornbill* (Oct-Dec 2004): 32–35.

Naoroji, R. & Punjabi, H. (eds) (2008) *Loke Wan Tho's Birds*. BNHS/OUP.

■ ■ ■

ABOUT THE BOMBAY NATURAL HISTORY SOCIETY

The BNHS was founded in 1883 and today it is the prime non-governmental conservation organisation in the Subcontinent. We work towards the conservation of nature and natural resources, education and research in natural history, and have members in over 20 countries.

Membership Activities and Benefits

- Nature camps to wildlife places both in and outside India.
- Treks, walks and field trips at weekends.
- Excellent audio-visuals presented by experts regularly.
- Seminars, workshops and correspondence courses on specific natural history subjects.
- Members receive *Hornbill*, a quarterly magazine.
- Subscription to the *Journal* is optional to members.
- Up to 15% discount on BNHS publications.
- 10% discount on BNHS products.
- Access to the finest collection of books on natural history.
- Voluntary Nature Education and Conservation activities.

Publications

BNHS Publications have been the standard reference works on the natural history of the Indian subcontinent since 1886. They are essential acquisitions for naturalists, amateurs and professionals throughout the country and abroad. Published uninterrupted since 1886, the *Journal of the Bombay Natural History Society* is acknowledged to be one of the finest scientific natural history sources for the Oriental Region. The popular quarterly magazine *Hornbill*, published since 1976, caters to a varied readership of all ages.

To become a member or for other details contact:

Bombay Natural History Society

Hornbill House, S.B. Singh Road, Mumbai 400 001, Maharashtra, India.

Tel.: +91-22-2282 1811 Fax: +91-22-2283 7615

Email: info@bnhs.org Website: www.bnhs.org

THE SOCIETY'S PUBLICATIONS

Notes